Manchester City Council Education Department

Technology For All

TEACHER FELLOWS

Helen Mount

Diana Ackerman

Manchester University School of Education

1991

David Fulton Publishers Ltd
2 Barbon Close, London WC1N 3JX

First published in Great Britain by
David Fulton Publishers, 1991

British Library Cataloguing in Publication Data

Mount, Helen
Technology for all.
I. Title II. Ackerman, Diane
607.1041

ISBN 1-85346-189-X

Designed and typeset by Vic Johnson, Manchester Education Department
Printed in Great Britain by Bell and Bain Ltd., Glasgow

Contents

List of Figures

Acknowledgements

This book could not have been published without the assistance and support of the teachers who allowed us to observe and write about activities undertaken in their classrooms and to whom we wish to extend our sincere thanks. We also appreciate being given the opportunity to share the learning experiences of those pupils involved, for without their co-operation our work could not have successfully been completed.

The headteachers of the six schools for pupils with severe and complex learning difficulties in Manchester have continued to give their support to our work and we are grateful for the commitment they have offered. We also wish to acknowledge the constructive comments offered by Professor Mittler, who has provided continued guidance whilst we have been based at the University of Manchester.

We would like to thank other colleagues within Manchester, particularly members of the Inspection and Advisory Service, who have read and commented on our work and finally Vic Johnson who is responsible for the page layout in this publication.

The case studies for Melland and Leacroft schools were written by Diana Ackerman and the case studies for Piper Hill and The Birches schools were written by Helen Mount.

Introduction

This book forms the sixth in a series which examines ways of providing meaningful access to the National Curriculum for pupils who have severe, complex[1] and profound and multiple learning difficulties (Fagg, Aherne, Skelton and Thornber 1990, and Ackerman and Mount 1991). The authors believe that the National Curriculum should be seen as a framework within which the present good practice in special schools across the country can be blended and developed. The concept of equal opportunities recognises that all children are entitled to experience a curriculum which offers them comparable quality and opportunity, irrespective of their level of attainment, beliefs, gender, social circumstances and the type of school they attend (Manchester City Council 1991a). It is against this background that Technology in the National Curriculum has been examined for pupils with severe and profound and multiple learning difficulties.

In talking to colleagues about Technology, many seem unsure about what it comprises, how it can be applied and how it differs from Science. Lever (1990) confirms this uncertainty and lack of confidence which some teachers feel with regard to Technology and discusses some possible reasons for this situation. One of the obstacles he found was the teachers' perceived lack of technological knowledge which reinforced their apprehension when tackling Technology in the classroom. This clearly has implications for in-service training, especially when he noted that access to Science and Technology courses was often given to the curriculum co-ordinator and/or the deputy head who, in most cases, were men. However, the majority of teachers in early years education (and schools for pupils with SLD) are women. Wyvill (in Browne ed.1991) reaches similar conclusions about the nature of Science which can be equally well applied to Technology. She suggests that the answer may lie in prejudices perpetuated in our society and emphasises ways in which women teachers can take action to improve their competence and make the subject more accessible to girls as well as boys.

From their research, Browne and Ross (in Browne ed. 1991) suggest that attitudes towards Science and Technology are firmly entrenched in children's' minds by the time they reach school, reflecting deep-rooted social patterns. They consider that attitudes cannot be modified merely by a single, simple strategy. However they stress that teachers within their classrooms can have some impact by promoting an atmosphere which fosters and nurtures equal opportunities by:

- developing positive attitudes towards various activities;
- developing confidence and choices;
- demonstrating careful consideration of the way in which activities are defined and identified;
- demonstrating careful consideration of the degree and nature of adult intervention and the methods of assessment.

The National Curriculum for Technology is not a subject which merely relies on acquisition of a particular body of knowledge. This should give confidence to those teachers who associate it traditionally with Craft, Design and Technology. It is much more wide ranging and concerns informed activity which might draw upon several different subjects and have several different outcomes. Five subjects, which until recently, were discrete and specialist curriculum areas have, for the first time, been encompassed by Technology in the National Curriculum. These are Art and Design, Business Education, Craft Design and Technology (CDT), Home Economics and Information Technology (IT). This amalgamation of subjects presents the class teacher with a chance to redress the balance in subjects which have previously been dominated by either girls (Home Economics) or boys (Craft, Design and Technology), thus promoting equal learning opportunities for all (Manchester City Council 1991b).

Pupils studying Technology will need to develop and apply particular skills, especially those of problem-solving, testing and evaluating, in order to be able to acquire and use knowledge about the Design and Technology process. The way pupils should acquire these skills and knowledge is through active learning in practical situations (DES 1987). In attempting to provide learning situations which enhance this process, it is essential to build on the way in which children learn about their environment. Initially this is by 'concrete' experiences - touching, pulling, grasping and tasting in order to investigate the nature of their surroundings. As children develop, so does their sphere of interest - from self, to family, the home, to schools and a wider society. They encounter more varied experiences and gradually absorb more complex situations (Mills 1991). The Non-Statutory Guidance for Design and Technology also recommends that pupils should begin by working in familiar contexts then progress to less familiar ones. In Key Stage 1, where most pupils with severe learning difficulties are likely to be working, it is suggested that any of the experiences offered will be in the contexts of the home, school and familiar aspects of the local community such as local shops. Pupils may also work in imaginary contexts in which there are few constraints and the programmes of study for Design and Technology in Key Stage 1 suggest that pictures, poems, stories and television programmes may provide starting points for some activities (DES 1990). Examples of these imaginary contexts are identified in the case study from Piper Hill School where stories and poems were used as a stimulus for technological activities.

Technology, like English, is concerned with activities that link with and draw on a wide range of curricular experience. It involves pupils in making many kinds of judgements - technical, social, aesthetic and economic, yet nevertheless it represents something distinctive and could be described as:

> ***"... the purposeful and practical response to perceived needs or opportunities, that must be worked out in particular contexts and against particular values. It is about being practically but intelligently capable - of doing, making, of knowing how. It cannot, therefore, be approached merely from books. Nor can its success be assessed in terms of the written word as opposed to activity..."***
>
> **(Pring 1989)**

The Non-Statutory Guidance for Technology in the National Curriculum focuses on the fact that it is a new subject and the Statutory Orders divide it into two profile components - Design and Technology capability and Information Technology (IT) capability. Both components emphasise the practical nature of the subject and the way in which it spans the curriculum, *"drawing on knowledge and understanding from a wide range of subjects"* (NCC 1990a). The Design and Technology profile component may be described as the investigation, development, planning, execution and evaluation of designs, whereas the IT profile component is concerned with the use of Information Technology across the curriculum. Both components are discussed in more detail below.

Lever (1990) suggests that there are two key issues to be addressed when approaching Technology:

- ***"...Technology is the application of skills for a purpose.***
- ***...Technology is NOT just linked to Science and/or C.D.T., but is essentially cross curricular and so can be utilised in ANY subject"***

(Lever 1990)

The programmes of study (PoS) have, in the past, provided a sound framework for access to the National Curriculum for pupils with severe and profound and multiple learning difficulties. They have outlined the *process* and the attainment targets (AT's) the *product* of pupil endeavour. With Technology, the arrangement of the PoS and AT's comprises a different format and requires careful and considered examination so that this group of pupils are given meaningful access. The Non-Statutory Guidance for Technology states that pupils should be allowed maximum participation whether or not they have special educational needs. Physical disabilities should not, of themselves, prevent pupils from engaging in design activities and if a pupil is unable to undertake a specific activity because of a disability, they may experience an alternative which closely matches it (NCC 1990a). The ways in which access can be provided will be outlined in more detail in the next section, which examines a thematic approach to the teaching and learning of Technology. This is followed by case studies describing current Technology practice in four of the six schools for pupils with severe and complex learning difficulties in Manchester.

Design and Technology

The PoS for Design and Technology suggest that pupils should design and make products which are categorised as:

- ARTEFACTS objects made by people;
- SYSTEMS sets of objects or activities which together perform a task;
- ENVIRONMENTS surroundings made or developed by people.

However, some products will fall into more than one category. For example, a car will be an artefact in that it is an object made for sale, having properties which can be compared with similar objects. It also comprises a series of systems which combine together to provide movement. It can finally be considered as an environment because it is a means of accommodating a person or group of people on a safe journey (NCC 1990a). During each key stage it is suggested that pupils should make each type of product.

In addition to developing and using artefacts, systems and environments the PoS throughout all the key stages are divided into three further sections. These are:

- working with materials;
- developing and communicating ideas;
- satisfying needs and addressing opportunities.

The Design and Technology profile component is essentially a process containing four attainment targets (AT's):

- AT 1 Identifying needs and opportunities;
- AT 2 Generating a design;
- AT 3 Planning and making;
- AT 4 Evaluating.

Feedback from each stage allows the pupil to either change or modify ideas already generated or to develop new ideas. It is also likely that some form of evaluation will take place during every stage. Design and Technology is the first area of experience in the National Curriculum where the attainment targets are a process in themselves and it is important to emphasise that this process does not necessarily start at AT1 and progress through to AT4 in a linear fashion:

> ***"The four attainment targets...define activities which are not intended to represent separable stages in a linear process. They feed both forward and back into each other and pupils' work towards one target will often relate to, and enhance performance on, others."***
>
> **(NCC 1989a para 2.4)**

Evaluation of an existing household appliance (AT4) could provide the stimulus for the design and technological process in which the identification of new needs and opportunities (AT1) would emerge. This may also generate new designs (AT2) which will then be planned, made (AT3), tested and evaluated (AT4). It has often been interpreted as a cyclical process, yet it is more than that. From the example just illustrated it is likely that the 'new' design generated from the original starting point may itself present opportunities for development, refinement or improvement when it is evaluated. As a result, the design process has entered a new phase which has more association with a spiral rather than a cyclical process because revisiting the four attainment targets is taking place at a better-informed and higher level of knowledge and experience (Fagg et al.1990a). Revisiting one area of the process in this example is providing information which a pupil can act on and apply to another area or attainment target.

The Non-Statutory Guidance for Design and Technology suggests that very young pupils or pupils with learning difficulties could be introduced to the subject through something which they would be capable of making. They would then investigate ways of improving it. The example provided shows that the process starts at AT 3 - Planning and Making (NCC 1990a), eg:

- Make a rubber-band powered car, using a cotton reel (AT3).
- Investigate which moving toys children prefer (AT 1).
- Decide on the toy to be made, eg a roundabout powered by a rubber band (AT2).
- Plan and make the toy (AT 3).
- Test how the roundabout works, and evaluate its usefulness as a toy (AT 4).

The Design and Technology research team for the Assessment Performance Unit (APU) identified two distinct aspects of capability that need to work alongside each other as pupils tackle a task. These are *reflective* (thinking around the task) and *active* (taking action on the task) (SEAC 1990). The two aspects of capability are linked in what they describe as an *"iterative process"* involving *"to-ing and fro-ing"* between thought and action. In order to explain this process more clearly, Figure 1 below shows an adapted version of the APU model incorporating the reflective and active aspects of Design and Technology capability.

Figure 1

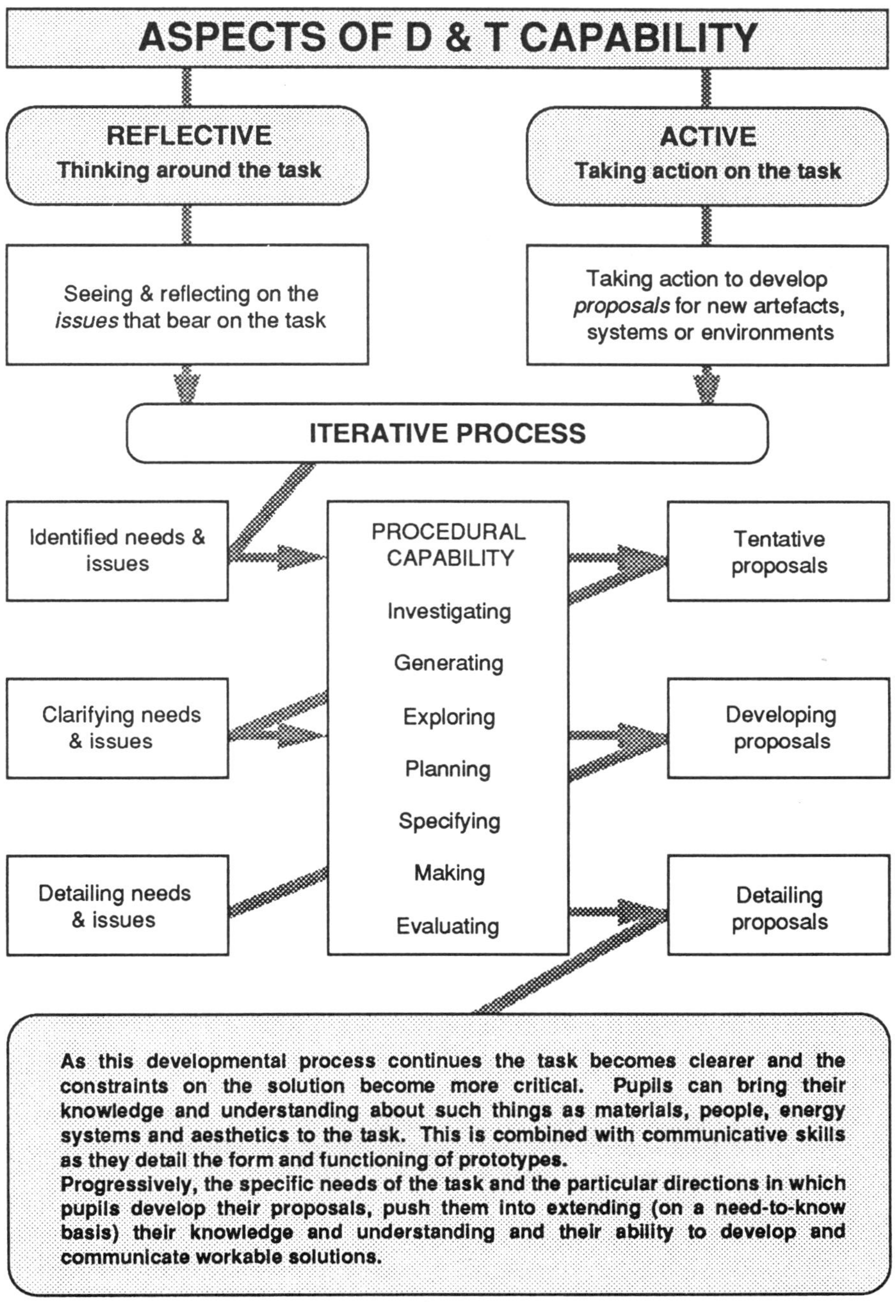

Capability in Design & Technology adapted from the APU Model (SEAC 1990)

Information Technology

Mention Information Technology (IT) and most people will immediately identify it with computers and/or word processors. The profile component for IT (attainment target 5), however goes much further than this. Information Technology is concerned with *"storing, processing and representing information by electronic means."* In addition to computers, it therefore includes microelectronic devices which are built into everyday objects such as cookers, microwaves, washing machines, cassette recorders, videos, etc. The PoS for the Information Technology profile component (AT 5) provide the minimum entitlement for all pupils and suggest that pupils working within level one should be taught:

- ***"that control is integral in many everyday products, such as cookers, cars, telephones;***
- ***how to give instructions to electronic devices, such as programmable toys and computers."***

Like the Design and Technology process, the attainment target for Information Technology is concerned with developing capability and using the skills acquired across the whole curriculum. Emphasis is given to:

- ***"Enjoyment and confidence in using IT.***
- ***Breadth of understanding of the use of IT in the world around us and the effects on daily life.***
- ***Encouraging a flexible approach so that children can easily adapt to change.***
- ***Enabling children to become familiar with the computer keyboard.***
- ***Enabling children to make choices about when and how to use IT in their work.***
- ***Developing perseverance"***

(Chadwick 1990)

The Non-Statutory Guidance for Information Technology suggests that pupils need to use IT in school to develop and enhance learning and to gain confidence and the capability to use IT in life after school (NCC 1990b). There is emphasis on the use of IT across the whole curriculum and the National Curriculum Statutory Orders for Science, Maths, English and Technology clearly indicate the contribution that IT can make to enrich the quality of learning (Manchester City Council Education Department 1990). HMI have also advised that:

> ***"All pupils should use a range of IT resources in core, other foundation and, where relevant, non-foundation subjects and cross-curricular themes. In the four key stages such resources might include electronic***

toys, calculators and musical instruments, as well as word processors and other computer software... Opportunities for using IT within the classroom are many and technological change continues to add to them and change them"

(DES 1989)

For many pupils with severe learning difficulties, Information Technology has provided the key to accessing much greater breadth and balance of learning opportunity during the last ten years in particular. This aspect is emphasised in the Non-Statutory Guidance:

"For pupils with communication problems or severe learning difficulties, information technology may provide the only viable means of access to the curriculum. For other pupils, information technology provides alternatives to text-based teaching; for example, synthesising music or designing with a computer"

(NCC 1990b)

The development of independence skills has been greatly enhanced by the simplification of control systems on electrical/electronic items such as cookers and washing machines. The use of communication aids with synthesised speech, concept keyboards and switches used in conjunction with specialised computer programs has also opened up possibilities for pupils with profound and multiple learning difficulties (NCET 1990). In addition, advances in technology have presented this group of pupils with greater access to leisure with the development of the 'Snoezelen' concept in Holland. Modified examples of this exist in a number of schools for pupils with severe learning difficulties in this country (Hulsegge and Verheuil 1987).

Pupils with severe and profound and multiple learning difficulties should have the opportunity to use IT for a variety of purposes:

- to communicate and handle information;
- to design, explore and evaluate;
- to use IT as a control mechanism;
- to make choices.

Uses of a computer may include:

- Identifying the main components of hardware - computer, monitor, printer, floppy disc.
- Switching on the above mentioned hardware.
- Handling and loading disc into the computer.
- Gaining access to the menu.
- Running software, use of keyboard, mouse, joystick or concept keyboard, word processing, data base programmes.
- Using a touch screen in conjunction with 'Touchstar' programme. Pupils are asked to touch the star, locate it on the screen to receive an audio-visual reward. A "Paint" programme also encourages the pupil to experiment moving her/his hand around the screen to create a picture of varying shapes and colour.

- Storage and retrieval of information, in which the pupils learn to save information on a disc to be retrieved when required .

(from the Melland School IT Curriculum 1990)

It is not the purpose of this publication to provide detailed analysis of the use of Information Technology with pupils who have severe and profound and multiple learning difficulties, as this warrants a book on its own. With regard to the National Curriculum, however, each of the four case studies contain specific reference to the use of IT within the chosen topic.

The next section discusses the basis for the case studies which are illustrated in the rest of this book. It shows how individual objectives may still be set and achieved by pupils with severe and profound and multiple learning difficulties within a thematic approach and that by blending the two together there is greater opportunity for interaction and learning. This is followed by four detailed case studies describing ongoing activities in four of the Manchester special schools. These activities have been blended primarily with the National Curriculum to illustrate that many of the learning experiences which are already being offered within the whole curriculum are compatible with the PoS for Technology as well as other foundation subjects.

[1]See Fagg et al. (1990) and Ackerman and Mount (1991) for further explanation of complex learning difficulties.

A Thematic Approach to Technology in the National Curriculum

Thematic or topic work is a well-established approach to teaching in many mainstream primary schools. It involves linking together a variety of often unrelated tasks and the associated activities are carried out under a common title or theme such as 'Colour' or 'Ourselves'. It is an approach which first gained widespread popularity in the 1960's after the Plowden Report made the observation that, *"Finding out has proved to be better for children than being taught"* (DES 1967).

It is not an approach without problems and teachers have, in the past, been criticised for merely offering superficial learning experiences in a limited number of subject areas. An HMI survey in 1978 expressed concern about the apparent lack of direction which was evident in some schools and the danger of children being exposed to unnecessary repetition unless considerable thought went into ensuring progression (DES 1978). Despite the difficulties, however, the thematic approach has outlived its critics and teachers in mainstream primary schools are still committed to this method which offers pupils opportunity to learn by discovering for themselves. Thematic work:

- ***"Crosses curriculum boundaries.***
- ***Involves practical activities.***
- ***Uses selected themes which are thought appropriate to the interests and stage of development of children involved.***
- ***Involves first hand experiences such as visits and visitors.***
- ***Involves using information gathering skills.***
- ***Includes class, group and individual work with some element of choice"***

(Bell 1990)

The advent of the National Curriculum has led many teachers of pupils with severe learning difficulties to examine their current practice and in recent years some schools or individual classes have also adopted a thematic approach. Whether this is compatible with the provision of individual learning experiences in an established objectives-based curriculum is one of several issues raised in an article by Byers

(1990). He suggests that reconciliation of the two approaches presents a challenge to teachers in special schools. It is, in his opinion, a challenge worth exploring because the advantages of both approaches can be capitalised on and the disadvantages avoided. Opportunities for broadening the special school curriculum and absorbing the requirements of the National Curriculum may be accomplished without presenting a threat to the traditional special school methodology. Examples given in the case studies will provide illustrations of how this can be achieved.

Within National Curriculum documentation for Key Stage 1, the importance of cross-curricular aspects is continually stressed. This is embodied in the core (cross-curricular) skills, themes and dimensions, and also in the guidance, programmes of study and attainment targets for the core and other foundation subjects. Collaborative group work which encourages peer led rather than teacher led interaction and experiential learning which encourages problem-solving, are both significant activities demanded of all children working within Key Stage 1. This is of particular relevance to Technology where the work is process oriented, concerned with not only 'what' children learn, but also 'how' they learn (Byers 1990).

The blending of thematic and objectives-based approaches is discussed in detail in Byers' article and a summary of the main features in the process provides a relevant background to the case studies described in this publication:

- Initially school staff (or class staff) choose a theme/topic and 'brainstorm' ideas which generates a topic web. The topic web comprises activities which can be differentiated to suit the needs of pupils of different ages and levels of achievement.
- Staff then relate the activities to the Whole Curriculum (not only the National Curriculum, but also to relevant school documentation). Cross-curricular aspects will become evident at this stage.
- Staff plan *"integrated schemes of work"* for individual pupils and groups of pupils. This involves -
  - (a) choosing age-appropriate and well differentiated activities.
  - (b) 'task analysing' those activities to create clearly defined objectives and activities for specific pupils.
  - (c) blending ongoing, high priority objectives through the topic-related experiences.

The four case studies which were observed over a half term or full term are described in the next section. They all took place within a thematic or topic-based framework and all were developed from existing practice being carried out in the classroom. The descriptions demonstrate that much of what is already going on is technological in nature and that the thematic approach can provide an appropriate forum for developing the technological process. Each case study outlines how the theme was developed, whether it was within a whole school or class and how individual and group objectives were addressed within it. Several sessions within each theme are highlighted in order to illustrate the differentiation, continuity and progression which took place. Ideas for activities which other practitioners will find useful are also offered. They include extended activities which facilitate further progression, either as an immediate continuation of the theme or when the theme is re-visited at a later date. Assessment and recording are an integral part of the approach and a selection of relevant individual and group records are illustrated. These form only part of the process, however, and other recording and assessment also includes pupils' own work, photographs and what they communicated (verbally, non-verbally) via audio and video recordings.

The case studies are illustrated in chronological order, starting with a class group of 7 - 10 year olds, followed by 11 - 14 year olds. The last two studies are devoted to senior groups of pupils aged 14 - 19 years and show distinctly different topics undertaken within the Technical Vocational Education Initiative (TVEI). In the interests of confidentiality, the names of the pupils and students have been changed. These case studies provide ideas which will help teachers to provide access to Technology in the National Curriculum for pupils with severe and profound and multiple learning difficulties. It is accepted that it is not always possible for teachers to know all the answers, but to approach work with *"Let's find out"* or *"Let's try this together"* is much more effective than *"Today, children we are going to make..."* (Lancashire County Council 1990). In observing and developing the activities in the case studies, we have witnessed pupils having fun and learning from their experiences in a positive and stimulating child-centred atmosphere.

Case Study 1 - Lower School: 'Trouble in the Ark' (Piper Hill)

Piper Hill is a group 2s day school for pupils with severe learning difficulties aged two to nineteen years. There are nine class groups and since the implementation of the Education Reform Act (1988) the school has been developing its curriculum and blending it with the National Curriculum.

Staff meet each term to discuss and agree a theme for the whole school and then 'brainstorm' to identify different areas which may be explored within the theme. The theme observed in this case study was 'Houses and Homes'. Figure 2 below illustrates some of the different areas of experience or topics which were highlighted in the initial 'brainstorming' session.

Figure 2

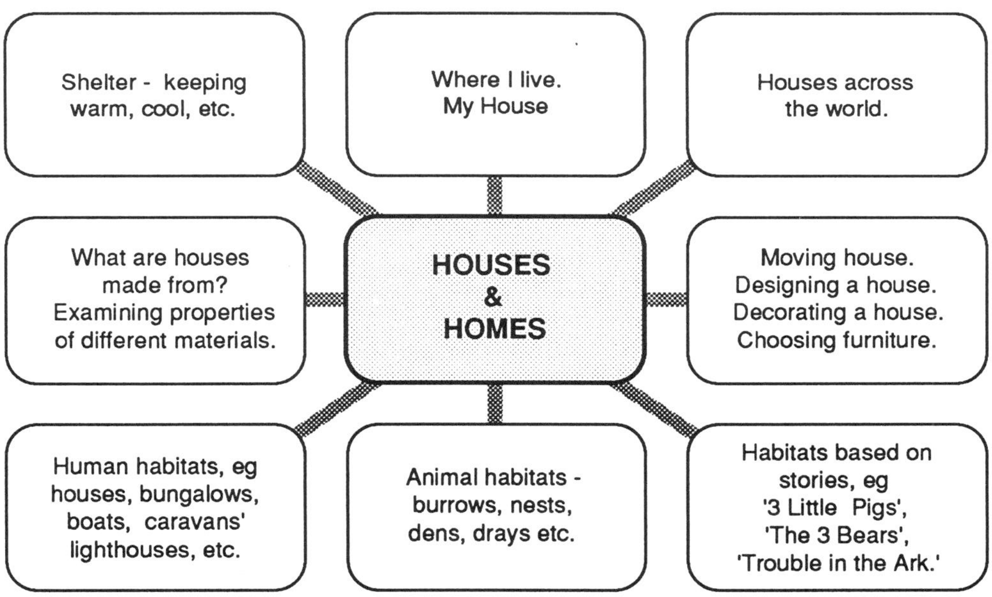

Suggested Activities for Termly Theme 'Houses & Homes'.

Once a selection of topics is identified within the general theme as outlined in Figure 1, staff choose one topic or several related topics which they consider most appropriate to the needs of pupils in their class, eg senior groups may choose 'animal habitats' whilst a junior class teacher may concentrate on a story as the starting point. The stories illustrated in Figure 2 are more suitable for infant/junior age pupils but could obviously be substituted with examples appropriate for seniors. Class staff then identify related areas of curriculum experience, together with ideas for activities which they record as part of their proposed scheme of work. Copies of this plan are inserted into the pupil's profile to provide a record of the breadth and depth of experiences. This ensures that pupils receive continuity and progression as well as breadth and balance in the same curriculum areas as they move through the school. If the proposed activities are not discussed and recorded it is quite possible for pupils to be exposed to the same experiences within the same topic year after year, thus narrowing their opportunity for learning.

The example chosen for this case study is one which developed from a story, *'Trouble in the Ark'* (Rose 1980) and took place in a class comprising nine pupils aged 7 - 10, three of whom have profound and multiple learning difficulties. The cross-curricular aspects of Design and Technology became evident as activities and experiences related to the topic were planned. Figure 3 is a modified version of the termly planning sheet which illustrates some activities and National Curriculum areas developed as a result of this class topic. It has been modified to highlight the close relationship between Design and Technology and other curriculum areas.

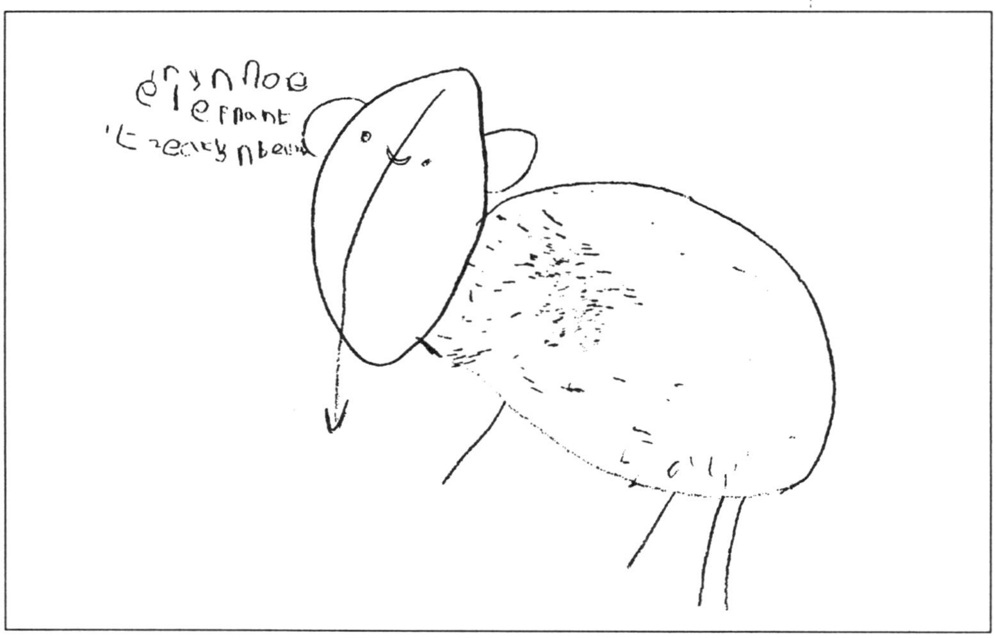

'An Elephant' by Brian, aged 7

The suggested activities are not meant to be exhaustive and it is not difficult to think of numerous other experiences which could be offered within the same topic. In the section on making books for example, pupils would be encouraged to explore different ways of making books to emphasise their story, eg on a word processor, making pop-up books, shape books, zig-zag books etc. They would be encouraged to justify their choice of materials and methods of display. In science, pupils would explore the properties of different materials and their effectiveness in keeping out the rain. They may then be required to make some protective clothing for Noah and his family or rainproof cover for a particular animal. Further ideas for activities are outlined in more detail at the end of this case study.

<table>
<tr><th colspan="4">Opportunities for Development of Cross-Curricular Links from a Story - 'Trouble in the Ark'</th></tr>
<tr><th>CURRICULUM AREA</th><th>ACTIVITIES</th><th>TECHNOLOGY</th><th>TECHNOLOGY SoA's</th></tr>
<tr><td>ENGLISH</td><td>Listening to and talking about the story.
Writing about a favourite animal.
Making a book about boats or animals.
Role-playing in the Ark built in class.
Discussing in all curriculum areas.</td><td rowspan="2"><u>AT 1 Level 1</u>
Describing why Noah may have needed an ark.

Suggesting features which would be required.

Describing what it would be like to live on a boat with so many animals.</td><td rowspan="2">"Pupils should be able to:

a) describe to others what they have noticed in familiar surroundings or visualised about imaginary situations.

b) suggest what might be done."</td></tr>
<tr><td>MATHS</td><td>Counting the animals 'two by two'.
Shapes in the Ark.
Sorting animals into sets.
Matching pairs of animals.
Measuring and estimating the sizes of different animals.</td></tr>
<tr><td>SCIENCE</td><td>Exploring floating and sinking with arks pupils have made.
Investigating the properties of materials.
Investigating animal body parts.</td><td rowspan="2"><u>AT 2 Level 1</u>
Explaining how they may start to build an ark and what features are essential.

Drawing pictures of their ark and giving reasons for their design.</td><td rowspan="2">"Pupils should be able to:

a) express their ideas about what they might do to meet an identified need or opportunity."</td></tr>
<tr><td>HISTORY</td><td>Recognising everyday time conventions, eg before, after, past, present, old, new.
Making comparisons between ancient and modern boats.
Examining the types of garments worn and utensils used.</td></tr>
<tr><td>GEOGRAPHY</td><td>Measuring rainfall.
Identifying land and sea on maps.
Transport - movement by boat.
Identifying different weather conditions.
Examining the effect of floods on the environment.</td><td rowspan="2"><u>AT 3 Level 1</u>
Using various materials to construct arks.

Using various materials to make masks for drama activities.

Making a rain gauge.

Making choices about which materials are likely to be the most suitable.</td><td rowspan="2">"Pupils should be able to:

a) use a variety of materials and equipment to make simple things."</td></tr>
<tr><td>ART</td><td>Making rubbings of animal skins/shells.
Making an ark wall frieze.
Making animals with stripes and spots.
Drawing and painting the ark.</td></tr>
<tr><td>DRAMA</td><td>Exploring ways of moving like one of Noah's animals and making animal noises.
Taking part in a play based on the book.</td><td rowspan="5"><u>AT 4 Level 1</u>
Describing to staff and other pupils what they have made and how well they have done.

Describing to others what they like or dislike about what they have made, eg mask, ark or wall frieze.

Discussing and evaluating properties of materials designed to protect people from extremes of weather, eg rain, snow, heat etc.</td><td rowspan="5">"Pupils should be able to:

a) describe to others what they have done and how well they have done it.

b) describe to others what they like and dislike about familiar artefacts, systems and environments."</td></tr>
<tr><td>P.E. / MOVEMENT</td><td>Developing animal movements and dances.</td></tr>
<tr><td>SOUNDS AND MUSIC</td><td>Listening to and making animal noises.
Listening to 'animal' music (classical).
Making sounds like the rain.
Singing related songs, eg The Animals went in Two by Two.</td></tr>
<tr><td>R.E.</td><td>Story of Noah in the Bible.</td></tr>
<tr><td>I.T.</td><td>Using relevant programs, eg 'Moving House'. Matching animal pairs.
Use of word processor to make book etc.</td></tr>
</table>

'Trouble In The Ark' - Cross-Curricular Aspects

Figure 3

Having identified curriculum areas and relevant activities within the topic, it is then possible to focus upon the way in which the educational opportunities were presented. It is also useful at this point to illustrate some of the group and individual objectives which were targeted within the topic. A selection of some of the half-termly objectives is outlined below and illustrates the way in which activities and objectives may be differentiated to take account of the needs of all pupils in the group.

GROUP OBJECTIVES (A sample only)

Whole Group	will develop inter-personal co-operation and wait their turn in activities, eg share construction equipment.
Robert, Brian, Jim, Carol, Elaine	will use a variety of materials to make arks under given criteria, eg - animals and people must fit through the entrance; will record their activities in writing, models and pictures (2D and 3D) will discuss and comment on each other's models.
Donna, Julie, Darren, Pam	will experience water activities using a variety of utensils and objects of different shapes, sizes and volumes; will develop fine motor skills by experiencing a variety of construction equipment.

INDIVIDUAL OBJECTIVES (A sample only)

Jim	will stay 'on task' for five minutes; will answer questions in context.
Robert	will construct at least one ark which floats and develop his understanding of materials and their uses.
Elaine	will draw a picture about one aspect of a story; will experience a variety of methods of construction and demonstrate and increase her confidence in experimenting.
Donna	will sit quietly during story time and take a fleeting interest in a book.
Brian	will comment without prompting on his own models and those made by others in the group.
Pam	will begin to explore construction equipment and other tactile materials with her hands and eyes rather than her mouth.

An example of a recording sheet for some of the group objectives is illustrated in Figure 4 and an individual record is shown in Figure 9 at the end of this case study.

Setting the Classroom Scene

At the beginning of this session the pupils were wearing face paints from a previous lesson. They had chosen one of the characters from the story and dramatised how they felt the animal would move and act. An 'ark' had been constructed in one corner of the room where it remained for pupils to use for role play and imaginative play. Wall displays contained a selection of animals with spots and stripes which had been made in art lessons. There were examples of pupils' writing and a wall frieze showed pairs

Piper Hill School **Group Recording Sheet**

Activity Using a selection of construction equipment to build an ark within school theme 'Houses and Homes.' Class topic - 'Trouble in the Ark.'	**Date** 22 April 1991 **Curriculum Area** Technology: AT 2 level 1 AT 3 level 1 AT 4 level 1a
Objectives To use Duplo to build an ark for the animals - giraffe, elephant, rhino. The animals must fit into the ark and must have a way of getting in, i.e. some kind of entrance. Pupils will record their work by drawing what they have made.	**Activity took place with** **PUPILS** Brian Robert Carol **STAFF** Class teacher

Evaluation

Initially not much interaction, but as arks were being constructed there was some curiosity about what others were doing and Robert helped Brian to find the piece he needed.

Robert thought about the height of the door. First, none of the animals would fit under his entrance (which looked like an arch) because it was too low. He made one side higher, then sides didn't match so couldn't put bridging piece on. Made sides level, but still not high enough for giraffe so dismantled and made sides higher still. The animals were then able to get into the ark and fitted in.

Brian talked about how his ark had to be high and pointed to the giraffe's long neck. Improved fine motor skills and problem solving - built up 2 bricks of small size to make the equivalent of a full one. Wanted to put a roof on his ark to keep the rain out, but the base was bigger than the piece he had so he joined two together.

Carol described very clearly what she was doing and talked about the fact that the animals needed to get into the ark before the rain came. She manipulated three large base pieces without joining them together or putting any other pieces on top so needed quite a lot of help to discover how to join them together.

Group Recording Sheet for One Session

Figure 4

of animals from the story. The language which pupils used when they were making the frieze had been recorded by the teacher and was illustrated next to the appropriate pair of animals or birds on the frieze, eg:

- "Orange rabbits - a tooth in its mouth." (Darren)
- "Our owls have got legs." (Jim)
- Robert thinks that Julie would say, "Two long ears, two legs, one tail, fluff, two rabbits."
- "Our owls have got eyes." (Brian)
- "Our ducks have got four legs." (Elaine)
- "Ducks have got tails." (Carol)

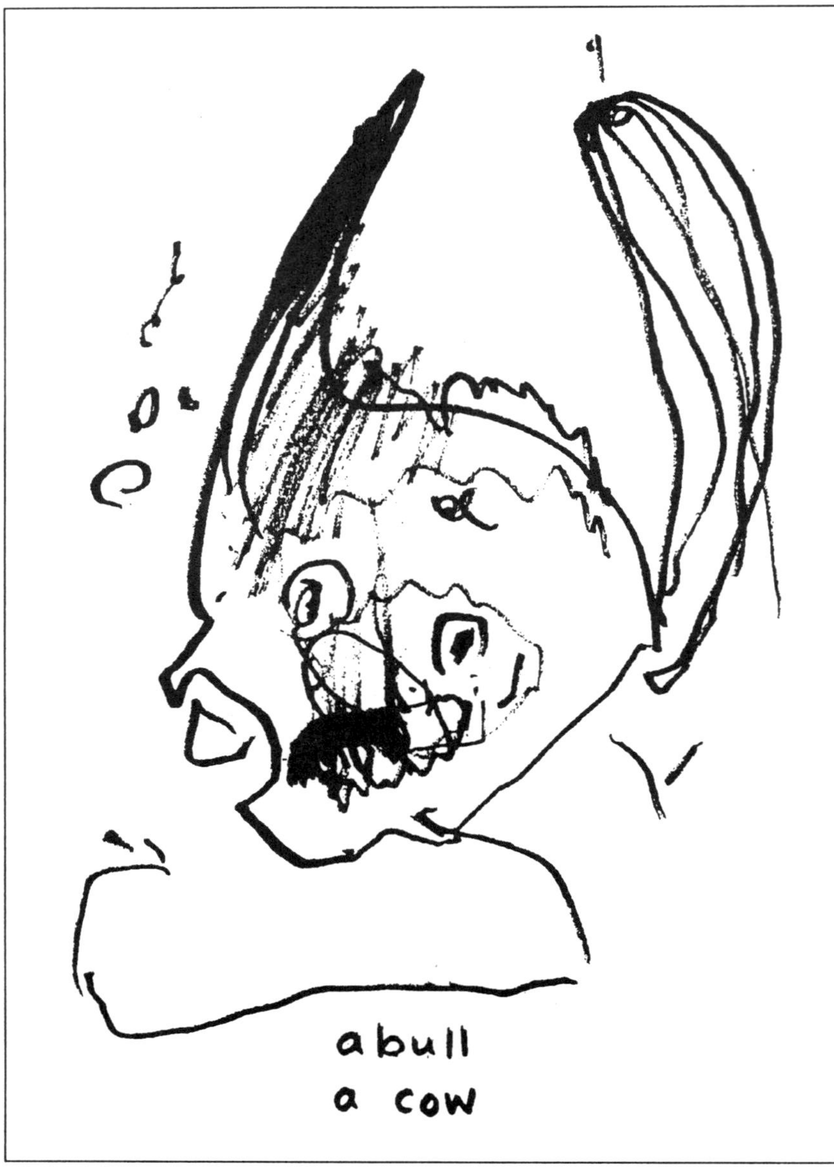

'The Bull' by Jim, aged 7

A sample of lessons was observed over a period of six weeks and these are summarised below.

Session 1

The whole class sat in a circle around the teacher who read 'Trouble in the Ark' from a 'big book' (Rose 1980). Pupils were encouraged to join in and it became clear that several pupils had remembered some of the story. They actively took part, filling in the end of a sentence or knowing which animal came next in the story sequence. The big book has an advantage in this mixed group of pupils as it assists those pupils who have a visual impairment to be able to see more clearly. Those pupils with profound and multiple learning difficulties appeared to show greater interest in the bright colours and clear pictures. One pupil who began by screaming, quietened as the teacher gave her a closer look at the book.

Through discussion during and at the end of the story, several of the pupils had identified the need for Noah and his animals to have an ark which would protect them from the rain and save them from drowning in the flood. The book had thus provided an initial stimulus for AT1 (Identifying needs and opportunities). Some of the pupils then decided that they would like to build an ark from Duplo (Lego), while others were given the opportunity to use large construction equipment on the floor to make their 'ark'. The pupils with PMLD, joined with another member of the class around a water trough containing various boats which was appropriate to the theme. They were later given a variety of construction equipment as an alternative experience.

In all the groups staff attempted to foster turn-taking, thinking and problem-solving skills so that the pupils could apply and extend their existing knowledge of the situation. In the Duplo group Robert Brian and Carol were given figures of Noah and a few animals which would have to fit into their ark. Robert was making his ark and it looked a little like a lid-less box. Initially, he had obviously not considered how the characters are going to fit in so Linda, the teacher intervened:

Linda: "How will Noah and the animals get into the ark?"

Robert: "Don't know".

Linda: "Think about it, will they all have to jump over the side of the ark? It might be too high."

Robert: "Need a door (to) get in."
(>>> Technology AT 1 - Identifying a need, AT 4 - Evaluation)

Robert then started to construct an entrance for the characters. At first it was much too low and he discovered that it needed to be higher. He modified it until he finally succeeded in his task and received obvious satisfaction, showing others that the characters went in through the entrance he had constructed.

Elaine, Jim and Shelley (a member of class staff) were on the floor nearby, constructing a large 'ark' from tubes of various lengths which had to be joined together with two, three or four-way connectors. Shelley was questioning Elaine about what sort of shape she wanted to make and why, as well as encouraging her to consider how to plan and execute her design:

Shelley: "What do you think we need to join these two pieces together?"

Elaine: "That one." (chooses correct connector).

Shelley: "Well let's see if it fits shall we? You do it."

Instead of letting Elaine know that she had selected the correct piece, Shelley allowed her to find out for herself. Elaine tried the two pieces together and they joined perfectly. She then had a little difficulty because one of her roof pieces was shorter than the other and would not reach the other side of the ark wall. Shelley encouraged Elaine and Jim to think of a solution together. Elaine initially attempted to find the piece of the correct length by trial and error. Shelley intervened again and prompted Elaine, helping her to discover (rather than be told) that the pieces were colour-coded and those of the same length were the same colour. Elaine was then able to select the correct piece by understanding more about the relationship between the length of the tube and its colour.

Throughout these activities the pupils were encouraged to discuss their own work - what they were doing, why they were doing it and how they might improve what they have done. They were also prompted to comment on each other's work, say what they liked or disliked about it and why, as well as offering suggestions to solve a particular problem.

At the end of the session the pupils displayed their Duplo arks on a table and predicted the buoyancy of their respective models. The teacher reminded them that their models would be tested at a later date when they had more arks to compare. The whole class came together to look at the results of each other's efforts and further discussion took place for the last few minutes of the lesson.

Figure 5 summarises the activities just outlined, together with the relevant Design and Technology programmes of study and attainment targets covered within this session.

Session 2

This session was a continuation of activities started in a previous one. The group were working towards presenting their class assembly on the story of Noah to the lower school. In previous sessions pupils had used face paints to provide them with their animal characters, but in a discussion it emerged that they would prefer to wear masks.

A wide selection of materials had been set out on tables around the room and small groups of pupils were busily engaged in activity and discussion with members of staff. The pupils with PMLD were able to join in what was effectively a whole class activity. Even though they could not verbally enter into the discussion, interaction between each member of the group was encouraged by the staff supervising them. After Pam (a pupil with PMLD) had been consulted about her choice by a member of staff, other pupils were encouraged to make suggestions about what sort of animal they thought Pam would choose if she could talk. They helped to choose materials for the mask and showed Pam what they had chosen, waiting for her 'approval'. She laughed and made fleeting eye contact with others, appearing to enjoy being engaged in activity close to the rest of the group. She was also able to explore the materials on the table, eg card, foil, paper, various fabrics, ribbon, tinsel, wood shavings, and was encouraged to inspect them visually instead of putting them straight into her mouth. With some physical prompting Pam picked up different materials, rubbed them between her fingers and held them gently against her face. This reinforced Pam's individual objective set at the beginning of the half term, which was to facilitate development from mouthing objects to exploring them with her hands and eyes.

The class teacher had started the session with a different book about Noah which was written as a poem and pupils were given a variety of stimuli to help them create their masks, eg large pictures and photographs together with templates of animal heads.

SESSION 1	Design & Technology : 'Trouble in the Ark'	
ACTIVITY	**D & T AT's**	**DESIGN & TECHNOLOGY PoS**
Involvement in the story - *Trouble in the Ark.* Writing about and drawing aspects of the story. Identifying the need for an Ark.	AT 1 Level 1	<u>Developing and communicating ideas</u> *Pupils should be taught to:* *• use imagination and their own experiences, to generate and explore ideas.* *• represent and develop ideas by drawings, models, talking,writing, working with materials.*
Discussing the choice of pieces and the way they might construct their arks. Discussing how they might construct a particular type of entrance making reference to the criteria set, eg how to make sure that it is high enough for the giraffes.	AT 1 Levels 1 and 2 AT 2 Levels 1 and 2	<u>Developing and using artefacts, systems and environments</u> *Pupils should be taught to:* *• know that a system is made of related parts which are combined for a purpose.* *• identify the jobs done by parts of the system.* *• give a sequence of instructions to produce a desired result.* *• recognise and make models of simple structures around them.* *• identify what should be done and the ways in which work should be organised.*
Building an ark from Duplo. Building an ark from other construction equipment.	AT 3 Levels 1 and 2	<u>Working with materials</u> *Pupils should be taught to:* *• explore and use a variety of materials to design and make things.* *• recognise that many materials are available and have different characteristics which make them appropriate for different tasks.* *• use materials and equipment safely.*
Evaluating their own construction and that of others in the group.	AT 4 Levels 1 and 2	<u>Satisfying needs and addressing opportunities</u> *Pupils should be taught to:* *• evaluate their finished work against the original intention.*

Design & Technology PoS & AT's Experienced in Session 1

Figure 5

There was considerable discussion throughout this session about many aspects of Design and Technology. Firstly, pupils had to remember which animal they had been in previous sessions and identify it in the story as well as select a picture or template of what the mask would look like. They then negotiated choice of appropriate materials for their chosen animal character. This included colour and type of decoration, eg whether to use paint or fabrics or a combination of both. Pupils also had to decide whether they felt the materials would be strong enough for the job which included trying different ways of joining things together - whether to use glue, sellotape or staples. Further discussion led to the discovery that the animal masks would need holes in the 'eyes' so that the pupils would be able to see where they were going while they were wearing them.

Once all the masks had been completed, the class group re-assembled to discuss their results. Masks were held up and the pupils guessed which animal it represented. They discussed the parts of the animal's face - whether it had nose, ears, eyes and a mouth and what they liked or disliked about it. Everyone liked Jim's lion mask which had a mane of gold paper strips and a brown painted face. However there was one critical observation from Robert:

Linda:	"What do you like about him?"
Elaine:	"Got gold." (>>> Technology AT 4 - Evaluation)
Linda:	"There's a name for this gold piece round his face. Can anyone remember what it's called?"
Robert	"Mane."
Linda	"Well done, Robert, That's quite a hard word to remember. Yes, it's called a mane ..."
Robert	"Got no ears." (>>> Technology AT's 4 & 1 - Evaluation & Identifying a Need)

The discussion which followed took a little time to resolve. Some of the pupils felt that the lion should have ears and others felt that they might be under the mane so could not be seen. They looked at some of the animal pictures used earlier in the session and finally agreed that the lion's ears were probably under its mane. There were no ears under the mane on Jim's mask which led to an evaluation of the purpose of the mask and whether it really needed ears as it was not a real lion.

'A Lion' by Robert, aged 10

For the last part of this session the pupils rehearsed their proposed assembly by acting the part of the animal they had characterised in their masks. Figure 6 summarises the activities which took place in this observed session in relation to Design and Technology. The activities were essentially art and craft, but the programmes of study and attainment targets covered, were similar to those outlined in Figure 5, and provide further illustration of the way in which Design and Technology permeates other curriculum areas.

Figure 6

SESSION 2	Design & Technology : 'Trouble in the Ark'	
ACTIVITY	**D & T AT's**	**DESIGN & TECHNOLOGY PoS**
Looking at animal pictures and discussing animal characters. **Tactile exploration of all the materials used.**	**AT 1 Levels 1 and 2**	**Developing and communicating ideas** *Pupils should be taught to:* *• use imagination and their own experiences, to generate and explore ideas.* *• represent and develop ideas by drawings, models, talking,writing, working with materials.* *• find out, sort, store and present information for use in designing and making.*
Deciding type of mask to make and what materials would be most appropriate. **Selecting material to make mask. Discussing and exchanging ideas about strength and aesthetics of materials chosen.**	**AT 2 Levels 1 and 2** **AT 3 Levels 1 and 2** **AT 4 Levels 1 and 2**	**Developing and using artefacts, systems and environments** *Pupils should be taught to:* *• make simple objects for a purpose.* *• know that a system is made of related parts which are combined for a purpose.* *• identify the jobs done by parts of the system.* *• give a sequence of instructions to produce a desired result.* *• recognise and make models of simple structures around them.* *• identify what should be done and the ways in which work should be organised.*
Exploring ways of joining pieces of mask together and ways of fastening mask to head or face. **Making the mask.**	**AT 3 Levels 1 and 2**	**Working with materials** *Pupils should be taught to:* *• explore and use a variety of materials to design and make things.* *• recognise that many materials are available and have different characteristics which make them appropriate for different tasks.* *• join materials and components in simple ways.* *• use materials and equipment safely.* *• choose materials and equipment to make objects.* *• investigate the properties of materials in the course of their designing and making.* *• identify natural and manufactured materials.*
Examining the work of others during the making stage and passing comment on the design of the mask. **Evaluating their own mask and those made by others in the group.**	**AT 4 Levels 1 and 2** **AT 4 Levels 1 and 2**	**Satisfying needs and addressing opportunities** *Pupils should be taught to:* *• realise that resources are limited and that choices must be made.* *• evaluate their finished work against the original intention.* *• ask people about their preferences.* *• recognise a variety of forms resulting from people's different values, cultures, beliefs and needs.* *• recognise aesthetic qualities in things around them and use them in their work.* *• talk about what they have learnt and what they might do differently next time.*

Design & Technology PoS & AT's Experienced During Mask-Making

Session 3

In this session Linda wanted to continue development of communication and problem solving skills through a sound and music session. The whole group took part as activities were just as relevant to the needs of those pupils with PMLD as they were to the rest of the group. Pupils sat around Linda at the beginning of the session and behind them were three tables containing a variety of musical instruments, eg xylophones, glockenspiel, tambours, tambourines, cabassa, etc.

The lesson started with *'Trouble in the Ark'* which was requested by several of the pupils and there was a lot of dialogue about the effects of the rain, how it caused the flood and how Noah protected his animals and family by building an ark. At the end of the story, Linda turned back to the pages where the rain started to fall and asked the pupils what happened to them when it rained. Robert, Carol and Elaine knew that they got wet and Linda then asked them if they could remember what sort of sound the rain made. Various vocal attempts were made by individual pupils who demonstrated their efforts to the rest of the class, moving close to the pupils with PMLD. This resulted in much hilarity as "plip plop" was chosen as the most popular phrase used at varying speed and pitch.

A sound effects tape of a thunderstorm was played and the pupils listened intently as the rain started quietly, then got louder with intervening thunder claps before it gradually faded away. Pupils were eager to listen to it a second time so it was repeated. After the tape had finished the group was asked to think of a different way of making sounds like rain - could they use different parts of their body rather than just their mouths? Robert clapped his hands, Elaine stamped her feet and other pupils were asked if they thought it sounded like rain. Linda intervened, suggesting that they could all try to make sounds like the rain on the tape, using some of the actions they had generated themselves. She asked them to follow her actions and started by gently and slowly tapping the palm of her left hand with one finger from her right. The speed and intensity increased in the following way:

- tapping more quickly with more fingers then moving on to clapping;
- clapping faster and faster then starting to slap legs;
- slapping legs quickly and increasingly louder;
- slapping legs and stamping feet and making the loudest sounds possible;
- gradually fading from the loudest sounds back to tapping gently with one finger and stopping.

Everyone thoroughly enjoyed this new activity and wanted to repeat it several times. Pam was very excited and laughed loudly as she was encouraged to move her hands and legs. Staff and other pupils also helped by patting their fingers and hands on Pam, Julie and Donna, to give them the tactile sensations that some of the other pupils could do for themselves. Some of the pupils were asked if they would like to demonstrate on their own and Carol immediately volunteered but the rest of the pupils said that it sounded more like heavy rain when everyone joined in together.

Linda had a sample of musical instruments behind her on a table which she brought into full view of the pupils. She asked them in turn to choose which one they thought they could make sound like rain. Once they had chosen their instrument, they tried them out, making soft and loud sounds or using the instrument in a less conventional way, eg scratching a tambour with their fingers rather than hitting it with a drumstick.

The class then split up into three small groups around the tables containing instruments and composed a sequence of sounds like rain which they later played back to the rest of the group.

This session was also preparation for the class assembly on the story of Noah and his ark. The pupils were working towards producing their own sound effects for the performance and composing their own song to complement the story. Design and Technology programmes of study were covered in this as in other sessions, but there was more emphasis on working within attainment targets one and four because pupils were mostly engaged in exploring familiar ways of making sounds and evaluating their success as a sound effect for their 'play'. In order to include the other two attainment targets for Design and Technology and to further develop the Design and Technology process, an extended activity could include pupils negotiating and making their own instrument to make rain sounds. They would be involved in discussing which of the available materials would be most suitable, planning their instrument, making it, testing it and evaluating it and recording their work.

Session 4

This was chosen as the final session in this case study because the pupils who had made arks in lessons over a period of two weeks were now at the point of testing them to see whether they floated or sank.

Pupils with PMLD had their own water tray and explored with a selection of plastic tubes of various diameters and lengths. Some of the containers had lids so they floated and others without lids eventually sank. Shelley demonstrated pouring water from one container to another and pouring water from a container back into the water tray. She spoke to the group and assisted the pupils to extend their action on the objects by using physical prompts.

The second group had constructed arks from Duplo, ReoClick (plastic construction material), plasticine and a selection of other materials, eg plastic tubes, plastic bottles, cups, card, paper, straws and cellophane. The arks made from these materials had been joined together in different ways and the pupils were eager to see whether they would float or sink. They were all set out on a table, clearly labelled with the pupil's name. Beside the water tray were two tables - one labelled 'float' and the other 'sink'.

The group sat round the water tray and Linda encouraged them to predict whether a selection of objects she had brought along would float or sink when put on to the water. Each item was tested in turn by the pupils who then put them on to the appropriate table. Some members of the group needed quite a lot of prompting from Linda and their friends at this stage. Pupils then tested their own constructions. Some of the Duplo and Reoclick arks floated and some sank. All of the plasticine models sank and all of the arks made from plastic tubes floated, but as the latter had been joined with sellotape, they began to disintegrate in the water after a short time. There was great excitement when an ark floated and disappointment when it sank (with a bit of humour!). Once all the arks had been tested and put onto the relevant float and sink tables, Linda selected a few samples to examine more closely and asked the group why they thought all the plasticine arks sank, why one of the Reoclick arks floated and the other sank (because the latter filled with water).

After the water was cleared away the group recorded the results of the trials by drawing pictures of the 'float' table and the 'sink' table with their respective contents. Whilst this was the final session observed for the case study the topic continued and there were

c

further extensions to the activities which had already taken place. Figure 7 provides a summary of the activities, illustrating how they relate to the programmes of study and attainment targets for Design and Technology at level 1 Statements of Attainment. The final column in Figure 7 shows the interrelationship between Design and Technology and the National Curriculum core subjects.

There has only been the opportunity in this case study to examine level one National Curriculum areas, but clearly some of the pupils had already achieved level one and were working within levels two and three of more than one of the Design and Technology attainment targets. An illustration of some of the ways in which Information Technology (AT 5) was used throughout the topic is provided in Figure 8. It highlights the work of three pupils who have different levels of attainment, illustrates some of the Information Technology experiences being offered within the National Curriculum and offers additional suggestions for further activities.

Even after the observations were over there was clearly considerable scope for extending much of the work carried out in previous weeks, especially with regard to the floating and sinking trials. Some ideas are outlined below:

- Continuing to work in a more structured way with the plasticine. The pupils' models were much too thick and heavy - some of them even had sails made from plasticine. It would, therefore be useful for pupils to explore ways of making a thinner structure which would also include the importance of getting an appropriate shape to try and make the ark float.

- Giving pupils further opportunity to experience making arks with a wider range of materials and tools for both making and joining, eg wood, metal, hammer, nails. The arks made from waste materials floated well initially, but as most of the pupils had used sellotape they quickly disintegrated, so there is obviously further exploration to be carried out with joining materials together. Linked with this is the purpose for which the finished article is going to be used and pupils could carry out plenty of investigations examining the waterproof nature of their 'joining' substances.

- Once a selection of arks had been developed, the testing could become more rigorous. They may, for example, be tested by putting 'animals' on board and observing how many it takes to sink the boat. If this is too difficult alternatives like counters could be used.

- Drawing conclusions about the most suitable materials for building an ark with given criteria and recording the results in a different way, eg using outlines of the animals to make a bar chart.

- Examining the efficiency of waterproof materials and discussing what sort of garment Mr. and Mrs. Noah would need if they had to go out on deck in the rain. Making a waterproof garment for a doll representing one of the characters. Evaluating the finished product.

- Visiting the Maritime Museum at Liverpool and/or going for a trip on the canal.

- Discussing the need for finding some way of measuring rainfall and eventually making and testing a rain gauge.

- Examining ways in which rain and resulting floods often cause damage.

SESSION 4	Design & Technology : 'Trouble in the Ark'		
ACTIVITY	**DESIGN & TECHNOLOGY PoS for Level 1**	**DESIGN & TECHNOLOGY SoA for Level 1**	**CORE CURRICULUM AREAS (SoA)**
Discussing whether objects presented would float or sink.	*"Use imagination, and their experiences to generate and explore ideas.*	Pupils should be able to: 1a) describe to others what they have noticed in familiar surroundings... 1b) suggest what might be done. (AT 1)	**ENGLISH** Involvement in discussion, asking questions, making suggestions. Recognising words 'float' and 'sink' placed on the relevant tables. Recording results of the experiment by drawings/ writing.
Choosing specific objects and predicting whether they would float and sink.	*Recognise that many materials are available and have different chacteristics which make them appropriate for different tasks.*	1a) express their ideas about what they might do to meet an identified need or opportunity. (AT 2)	
Testing whether objects would float or sink.	*Use materials and equipment safely.*		**MATHS** Using materials provided for a task. Talking about own work and asking questions. Making predictions based on experience. Estimating how many objects floated and how may sank. Sorting 3D and 2D shapes. Building with 3D shapes, drawing 2D shapes and describing them. Recording results and commenting.
Discussing results and making suggestions why something happened, eg ark floated or sank.	*Describe to others what they have noticed.* *Talk about what they have done during their designing and making.*	1a) use a variety of equipment and materials to make things. (AT 3)	
Discussing why it was important for Noah's ark to float.	*Evaluate familiar things by observing and describing them, saying what they like or dislike about them and why people have or need them.*	1a) describe to others what they have done and how well they have done it. 1b) describe to others what they like and dislike about familiar artefacts, systems and environments. (AT 4)	
Testing pupils' own constructions and recognising that they either floated or sank.	*Evaluate their finished work against the original intention.*		**SCIENCE** Observing familiar materials and events at first hand, using their senses. Describing and communicating their observations in groups within their class. Knowing that there is a wide variety of living things. Being able to name body parts (of the animals). Describing familiar and unfamiliar objects in terms of simple properties, eg size, shape, texture. Knowing that there is a variety of weather conditions. Being able to describe changes in the weather.
Putting objects which floated and sank on to the correct table.	*Represent and develop ideas by drawings, models, talking, writing, working with materials.*	(DES Technology 1990)	
Recording the results by drawing the objects which floated and sank as sets.	*Find out, sort, store and present information for use in designing and making."* *(DES Technology 1990)*		

Summary of Testing & Evaluating Arks Showing Cross-Curricular Links

Figure 7

Figure 8

Piper Hill School **Sample Group Assessment Sheet** **SUBJECT / AREA** Information Technology

AT	LEVEL	ACTIVITY	PROGRAMME OF STUDY Pupils should be taught:	WORKING WITHIN SoA Pupils should be able to:	OBSERVATIONS
5	1b	PAM : purposefully used a single switch to activate the tape recorder during the music lesson.	how to give instructions to electronic devices.	talk about ways in which equipment responds to signals.	Pam is beginning to use the single switch to cause change and make choices, eg activating lights, music.
5	2a	BRIAN : chose to work on the programme "Moving House." He invited Carol to join him in the game.	that information can be used to help plan and organise ideas in written form.	use computer-generated pictures, symbols, words and phrases to communicate meaning.	Brian can access the computer by himself. He can also use up to four keys on the keyboard to access / operate program.
5	1a	ROBERT : played an animal matching game on the computer, using the concept keyboard.	to know that information can be held in a variety of forms.	work with a computer.	Robert enjoys using the concept keyboard with his own words acting as the word bank. He is able to build simple three word sentences, by pressing the overlay.

ADDITIONAL IDEAS:

- Using the concept keyboard to play a variety of matching, sorting animal games.
- Using electronic organ to create loud / soft sounds for weather eg - soft, quick sound for rain.
- Using tape recorder to listen to taped animal noises and record player to hear favourite records.
- Using computer and word processor to type and print own stories, record children's favourite animals, build up a word bank that can be saved on disc and retrieved, when required.

- Discovering that despite some of the problems, water is essential for various life forms and without it they would die. Experiments with plants would provide a useful starting point for this activity which could be extended to an examination of other essentials for life, eg nutrient, warmth, the sun.

Recording

A variety of methods was used to record pupil progress through the activities within the topic at Piper Hill. Pupils' work in the form of drawings, paintings, pictures and writing were selected to go into their Records of Achievement. This was supplemented by the use of photographs and video tape of pupils engaged in different activities, which is a particularly useful method of recording and assessing the progress of those pupils with profound and multiple learning difficulties.

In addition, the class teacher kept an ongoing record of both group and individual responses to the activities in relation to the objectives which had been set at the beginning of the topic. An example of an individual recording sheet is illustrated in Figure 9.

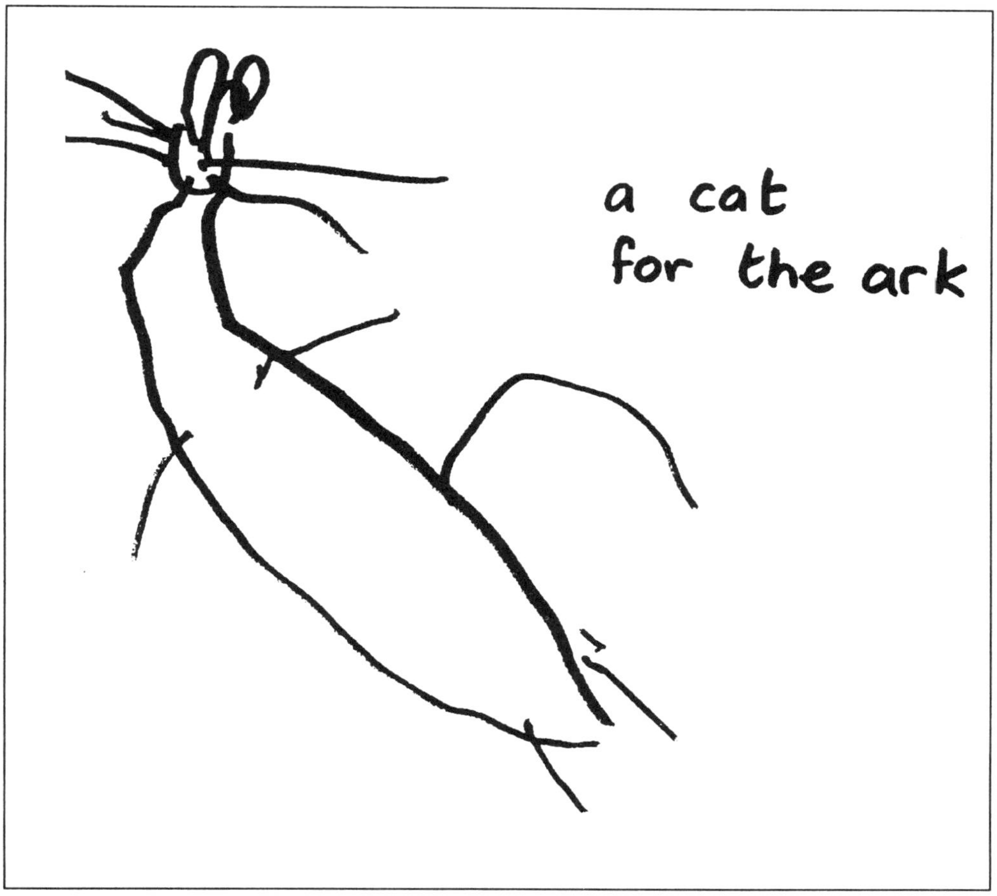

'A Cat for the Ark' by Jim

Individual Recording Sheet

CURRICULUM AREA Technology

NAME Robert

AIM
- To explore & use a variety of materials to design and make things within the theme.
- To apply and develop skills learnt.

DATE SET: April 1991

TOPIC 'Trouble in the Ark'

DATE TO REASSESS: June 1991

NATIONAL CURRICULUM: AT1 1a & 1b; AT 2 1a; AT 3 1b

WHOLE CURRICULUM: English, Maths, Science, Geography, History, Music, P.E., Art, PSE.

DATE	ACTIVITY	EVALUATION
22 April	Story, then construction using Duplo	Set off making and ark without really considering that animals and people had to fit in. Realised errors and changed structure.
26 April	Construction - Reoclick	More concentration today on criteria. Made opening door. Wanted it to "swing".
1 May	Mask - making	Selected grey fabric to make trunk for elephant and stuck it on with glue. "Need holes to see" (where he was going).
7 May	Construction using Duplo	More interaction with peers noticeable. Worked with Elaine and told her what to do. "No, not this brick, that one."
10 May	Construction using Duplo	Continuation of previous work with Elaine. Gave very clear instructions. If she did not understand, he showed her what he meant.
17 May	Construction using 'junk'	Made ark from plastic tubes. Realised he would have to find lids to block ends otherwise it would sink - "fill with water, go down."
24 May	Construction - plasticine	Left to explore and made ark that looked like a raft. He was confident that it would float when tested.
24 May	Testing - floating/sinking properties of arks	'Junk' ark floated and he was delighted, but he could not understand why his plasticine ark sank at first. Eventually said "Too heavy".

Individual Recording Sheet for Part of Topic.

Figure 9

Case Study 2 - Middle School: 'Growing Vegetable Soup' (Melland)

Melland school is a group 2s day school for pupils with severe and profound learning difficulties. With the advent of the National Curriculum, staff at Melland have sought to mesh its existing objectives-based school curriculum with the more interactive, open-ended requirements of the National Curriculum.

In the spring of 1991, a small garden plot in a nearby college was made available to the school. One of the teachers, working with a class of 12-14 year olds decided that "Growth and Change" would be an appropriate class theme for the spring and summer terms. He chose to concentrate on the topic area of 'vegetables' and devised a set of activities that would cover both terms, allowing time for growing, tending and harvesting the vegetables. Figure 10 below illustrates the class topic web and explores some of the ways in which this theme could be developed.

Figure 10

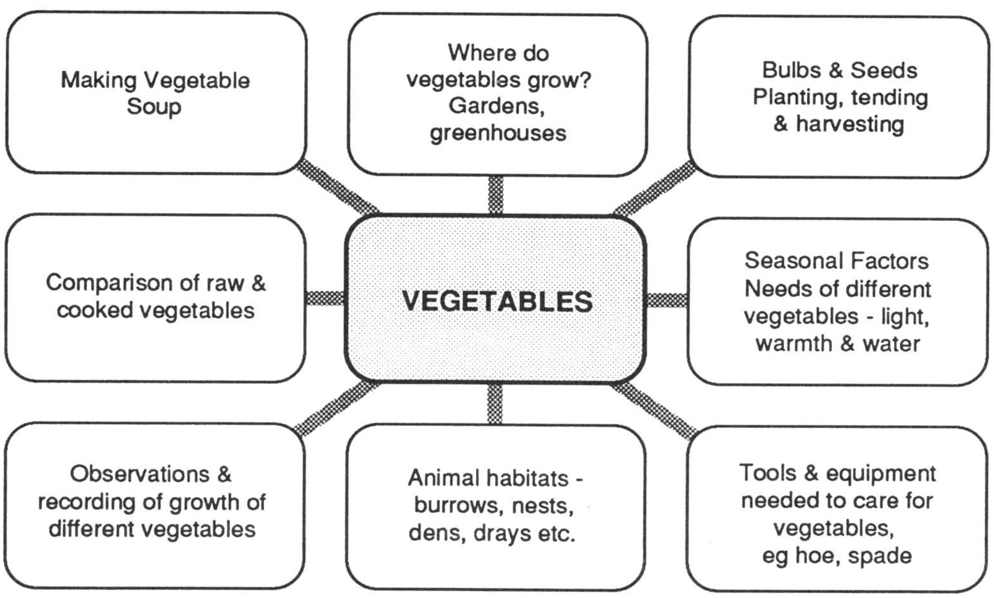

A Sample of Suggested Activities for Topic of Vegetables

The teacher, then related the activities to the National Curriculum subject areas and the school's own curriculum documents emphasising the cross curricular, whole curriculum nature of the activities and incorporating core (cross curricular) skills, themes and dimensions. Figure 11 illustrates the inter-relationship between Technology and a number of core and foundation subjects.

It had also become clear that the Design and Technological aspects of the topic would be embedded in the pupil's work within many related curriculum areas. This was highlighted in the creation of the class vegetable garden. Figure 12 illustrates the pupils' experiences of Design and Technology within the topic, "Vegetable Soup."

The teacher devised a series of lesson plans which would incorporate activities for both individual pupil needs and group objectives for his mixed ability group. The group activity would be open-ended allowing for collaborative work and co-operation between pupils, and at the same provide the opportunity to differentiate activities within it, i.e. an Integrated Scheme of Work. Examples of these can be found within some of the lesson plans. Following is a sample of lessons observed in the latter part of the summer term in which some of the aims for the whole group were as follows:

- to be introduced to Design and Technology capability by exploring a familiar situation, "The garden";
- to draw up plans for a vegetable garden and create a recipe book for vegetable soup;
- to identify the range of items commonly found in a garden;
- to explore and use a variety of materials and select appropriate materials for a given task;
- to be encouraged to represent ideas by drawings, models, talking and writing;
- to use simple tools safely and appropriately;
- to experience the breadth of understanding of the use of IT.

During the spring term, the pupils had planned and designed their vegetable garden, selected and planted the vegetable seeds and plants and were ready to harvest the vegetables and make their home-made soup.

'Planting the Vegetables'

<table>
<tr><th colspan="4">'Vegetable Soup' : Cross-Curricular Aspects</th></tr>
<tr><th>CURRICULUM AREA</th><th>ACTIVITIES</th><th>TECHNOLOGY</th><th>SKILLS, THEMES & DIMENSIONS</th></tr>
<tr><td>ENGLISH</td><td>Speaking and listening in discussion groups about vegetables
Reading and writing own class recipe book
Reading the story "Vegetable soup."</td><td rowspan="10"><u>AT1 Level 1</u>

Describing why they need a recipe book.

Suggesting different ways of making soup.

Describing what it might look like when it is finished.

<u>AT 2 Level 1</u>

Sharing their ideas about how to design a recipe book.

<u>AT 3 Level 1</u>

Using a variety of tools to prepare the vegetables.

Making the recipe book.

Making choices about which vegetables are suitable for soup.

<u>AT 4 Level 1</u>

Describing to staff and other pupils what they have made. Commenting on their own contribution and that of others.

Discussing with others what they like or dislike about what they have made.

Evaluating the " veg soup" and suggesting improvements or changes to improve the taste, appearance of the soup.

<u>AT 5 IT</u>

Using a computer to record and label .</td><td rowspan="10"><u>Skills</u>

Communication.

Numeracy.

Problem solving.

Information technology.

Recording.

Researching.

Planning.

Inter-personal, ie listening negotiating, questioning, talking etc.

<u>Themes</u>

Environmental awareness.

Health and safety.

Simple economics of supply and demand of food.

Economic and industrial awareness, eg need for vegetables to have light, warmth and water to grow

<u>Dimensions</u>

Equal opportunities in experience offered to girls and boys.

Raising awareness of other cultures through tasting food from other countries eg In China, soup may follow the main dish.</td></tr>
<tr><td>MATHS</td><td>Sorting fruits and vegetables into sets
Examining different shapes of both
Similarities and differences in size
Estimation of random number of vegetables
Cutting vegetables into halves, quarters
Addition and subtraction using vegetables
Making a block graph of vegetables - children's likes and dislikes.</td></tr>
<tr><td>SCIENCE</td><td>Observing changes in vegs. when cooked
Cutting slices of veg. and match to whole vegetable
Ways of keeping vegetables fresh
Tactile and tasting experiences
Healthy and unhealthy food
Effect of weather on vegetable growth.</td></tr>
<tr><td>HISTORY</td><td>Recognising everyday time conventions, eg before, after, past, present, old, new.
Making a time chart which shows the growth of a plant over a period of time.</td></tr>
<tr><td>ART</td><td>Vegetable prints on a variety of materials
Large wall frieze depicting a vegetable garden containing root and green vegetables
Illustrating the cover of recipe book
Using a variety of modelling materials to make models of vegetables.</td></tr>
<tr><td>GEOGRAPHY</td><td>Children create their own garden map keeping track of weather and seasons.</td></tr>
<tr><td>P.E. / MOVEMENT</td><td>Growth of a plant,
Actions associated with gardening, eg digging, planting, raking and hoeing.</td></tr>
<tr><td>SOUNDS AND MUSIC</td><td>Listening to music eg "The Four Seasons."
Using percussion instruments to illustrate rain, wind, thunder
Variety of songs.</td></tr>
<tr><td>R.E.</td><td>The importance of caring for living things.</td></tr>
<tr><td>I.T.</td><td>Labelling pictures
Using a word processor to make a recipe book.</td></tr>
</table>

Figure 11

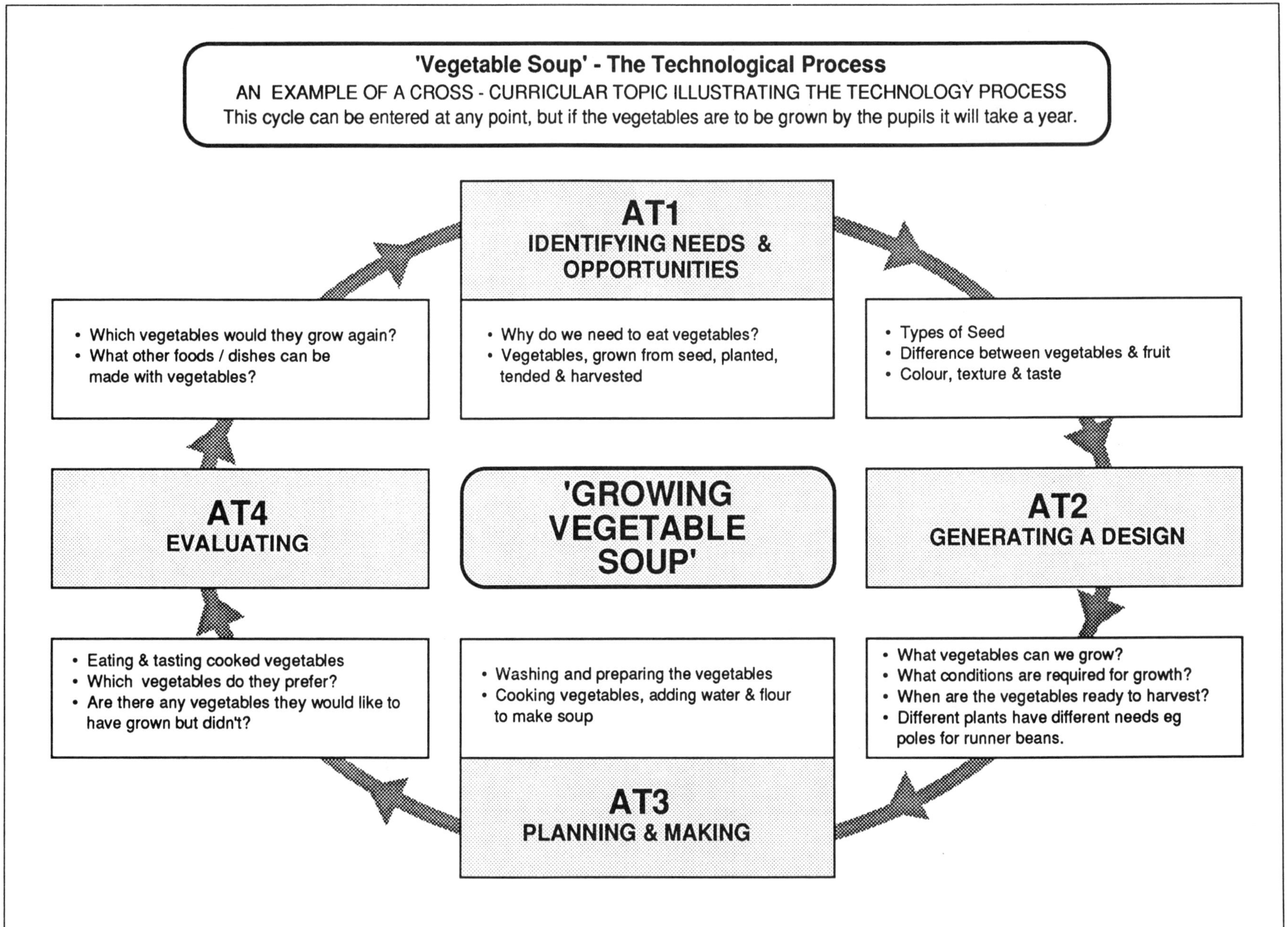
'Vegetable Soup' - The Technological Process
AN EXAMPLE OF A CROSS - CURRICULAR TOPIC ILLUSTRATING THE TECHNOLOGY PROCESS
This cycle can be entered at any point, but if the vegetables are to be grown by the pupils it will take a year.
AT1
IDENTIFYING NEEDS & OPPORTUNITIES
• Why do we need to eat vegetables?
• Vegetables, grown from seed, planted, tended & harvested
• Types of Seed
• Difference between vegetables & fruit
• Colour, texture & taste
AT2
GENERATING A DESIGN
• What vegetables can we grow?
• What conditions are required for growth?
• When are the vegetables ready to harvest?
• Different plants have different needs eg poles for runner beans.
'GROWING VEGETABLE SOUP'
• Washing and preparing the vegetables
• Cooking vegetables, adding water & flour to make soup
AT3
PLANNING & MAKING
• Eating & tasting cooked vegetables
• Which vegetables do they prefer?
• Are there any vegetables they would like to have grown but didn't?
AT4
EVALUATING
• Which vegetables would they grow again?
• What other foods / dishes can be made with vegetables?

Figure 12

The classroom reflected the range of activities that had been covered during the previous term. A large wall frieze, made during an art lesson, distinguished root from green vegetables and a selection of vegetables were displayed on a nearby table. A sorting tray was close at hand with pictures of individual vegetables placed in the compartments of the tray.

The following ongoing science activities with seeds were also observed:

- seeds planted in various media (monitored on a graph);
- seeds kept at different temperatures;
- seeds planted at different depths;
- seeds planted close together and far apart;
- seeds receiving different amounts of water.

The children had also experimented with growing different parts of plants, eg carrot tops and potato eyes and had made a variety of salt dough model vegetables, painted, varnished and set out on a nearby display table. Pupils' likes and dislikes of different vegetables were recorded on a block graph on the wall.

Sample Lesson 1 : 'Harvesting the Vegetables'

The children were very excited about harvesting their vegetables towards the end of the summer term. It represented many weeks of hard work - digging, hoeing, preparing the soil, planting, weeding and tending the plants. All the children had taken part in this weekly activity. Once they arrived at the garden there was lots of discussion about the different jobs that would need doing. Each child was encouraged to assume

'Harvesting the Vegetables'

responsibility for at least one job. This involved considerable negotiation, before the jobs were distributed as follows:

- James and Peter had chosen to dig up the potatoes.
- Carl had decided he would harvest the cabbages and he asked if Shazad could help him.
- Elizabeth had chosen to collect her favourite vegetable, carrots and wanted Noreen to hold the basket.
- Greg, who uses a wheelchair and Don (the nursery nurse) were ready to harvest the onions from their raised plot. The raised beds allowed the pupils with profound and multiple learning difficulties to be participants and not just onlookers in the garden plot activities.

When the pupils arrived at the garden, Keith (the teacher) encouraged them to think about what tools they would need to harvest the vegetables and how they would remove them from the ground. The children tried different tools to discover for themselves which ones were the most appropriate for extracting the different vegetables. Elizabeth thought she could pull up the carrots with her hands, but after repeatedly pulling up handfuls of carrot leaves and no carrots, she decided to use a small fork and dug deeply until she could grasp a carrot and remove it intact.

James and Carl were unsure about how to remove the potatoes and asked Keith for help. He encouraged them to try on their own first, offering support and further encouragement until James realised that they needed a big fork to dig deeply to retrieve the potatoes. Keith showed them how to loosen the soil and pull the plant gently by its leaves so the leaves and plants were removed together. Peter noticed that some of the leaves had withered away and James pointed to the potatoes that were mushy and black. The teacher noted that these observations could form the basis for further activities on pests, diseases, decay and waste.

Peter quickly worked out that they could pull up the cabbages with their hands and demonstrated this to Shazad. Keith encouraged the children to examine the roots of each vegetable and compare them with those of other vegetables. Carl had observed that there were big holes in some of the cabbage leaves and there was some discussion about them. Keith reminded him that that on a previous visit, they had seen some caterpillars on the leaves.

Greg and Don were working hard to harvest the onions, shaking each one to get rid of the soil. James, who was working close by pointed out the purple flowers on one or two of the onions and couldn't understand why there were flowers on the vegetable plants. Keith explained that if the onions were left in the ground for much longer, they would all grow flowers. Elizabeth offered to help pull up some onions, but pulled too hard and the stem snapped off. Immediately, the children could smell the onion and Greg turned his head away to avoid the smell. Figure 13 summarises the differentiated technological activities and cross-curricular nature of this sample lesson which promoted interaction, collaboration and negotiation in a meaningful, functional context.

The children worked together, guessing what vegetable they were trying to dig up, looking carefully at the leaves of the individual plants in order to identify them. Keith and Don brought Noreen onto a blanket on the ground nearby, where she could touch and smell the soil and the vegetables. Elizabeth accidentally touched a worm and there was some discussion about the usefulness of worms, but she remained unconvinced and eventually buried it under a mound of soil. When all the vegetables had been harvested, the teacher again drew the children together and finished the session by encouraging them to close their eyes and guess the vegetable by feeling the shape or by the smell. Finally, the vegetables were sorted into groups and packed

Melland School SAMPLE LESSON 1			
Individual & Group Objectives within an Integrated Scheme of Work 'Harvesting vegetables for homemade vegetable soup'			
		Technology PoS ***Pupils should be taught that:***	**Pupils should be able to:**
James	will record in a diary the number of vegetables harvested **Maths, Technology, English**	*IT can be used to store information*	**work with a computer** **AT5 SoA 1a**
Noreen	will hold basket on lap (in wheelchair) to receive harvested green vegetables **Self-help**	*materials have different characteristics that make them appropriate for different tasks*	**express their ideas to meet an identified need** **AT2 SoA 1a**
Carl	will use a trowel and spade to dig up carrots and potatoes **English, Maths, Geography**	*use materials and equipment safely*	**show that they can use simple hand tools** **AT3 SoA 2c**
Shazad	will pull up some vegetables (unaided) and place in basket on the ground **English**	*find ways in which work should be organised*	**suggest what might be done** **AT1 SoA 1b**
Elizabeth	will identify ripe and unripe vegetables and choose to pick ripe vegetables **Science**	*evaluate familiar things by observing them*	**Describe to others what they have done and how well they have done it** **AT4 SoA 1a**
Greg	will feel a variety of vegetables to enhance tactile awareness Monitor likes and dislikes **Sensory and Science**	*describe to others what they like and dislike*	**Describe to others what they like and dislike about familiar artefacts systems or environments** **AT4 SoA 1b**
Peter	will sort vegetables into two separate baskets • root vegetables • green vegetables **Maths, Science**	*use their own experience to explore ideas*	**Describe to others how they are going about their work** **AT3 SoA 2a**
Whole group	• will harvest the vegetables in preparation for making vegetable soup • will use a variety of gardening tools to complete the task • will wash and dry their hands after harvesting **PSE self- help**	*• talk about what they are doing* *• explore and use a variety of materials and equipment* *• use materials and equipment safely*	**Discuss with teacher how satisfied they are with their D & T activities, taking into account their original intention and how they went about their task** **AT4 SoA 2a**

Figure 13

into muslin bags. The children enjoyed the session, had practised many newly acquired skills, explored many concepts and experienced the satisfaction of reaping the rewards of their hard work.

'Tending the Crops'

Figure 14 provides a summary of the activities in the garden, incorporating the programmes of study and the statements of attainment within levels 1 and 2 for Technology, which have then been referenced to the National Curriculum core subjects.

Sample Lesson 2 : 'Making a Recipe Book'

The whole class sat in a semi-circle around the teacher and discussed the need for a recipe book for the vegetable soup they were planning to make. James had found a book, "Veg. Soup" at his local library and shared it with his classmates but had to return it to the library. It became evident that the children would need to make their own recipe book. A display table containing a variety of vegetables was nearby. Two tables covered with a variety of creative materials, scissors, felt tips, paper, card of different sizes and glue were close by. On a small table near the sink were some bowls of coloured poster paints and some pieces of vegetables, which the children had cut up earlier in the day. A computer and printer were set up in the corner of the room and Figure 15 is an illustration of a sample of Information Technology activities, that were available to the children.

There was an animated discussion about the sorts of vegetables that should be included and the teacher encouraged all the children to take part in the conversation. When Carl suggested "onions" Elizabeth tried to shout him down but the teacher gently reminded her that this book was a class book and everyone should be able to include their own favourite vegetable. The children constantly touched or pointed to

Sample Lesson 1 | **Design & Technology : 'HARVESTING VEGETABLES'**

ACTIVITY	D & T AT'S	DESIGN & TECHNOLOGY PoS Pupils should be taught to:	DESIGN & TECHNOLOGY SoA's Pupils should be able to	RELATED CORE CURRICULUM AREAS
Describe the different vegetables in the school garden. Offer suggestions about how to protect the vegetables.	**AT 1 Level 1 & 2**	*identify what should be done and ways in which the work is organised.*	**Describe to others what they have noticed.** **Suggest what might be done.**	**MATHS** Using materials provided for a task Talking about own work and asking questions. Making predictions Comparing and ordering objects without measuring. Giving a sensible estimation of a small number of objects.
Discuss different ways of harvesting the vegetables. Share ideas about possible food dishes that could be made with the vegetables.	**AT 2 Level 1 & 2**	*use imagination and their own experience to generate and explore ideas.* *represent and develop ideas by drawings, models, talking, writing, working with materials.*	**Express their ideas about what they might need to meet an identified need.** **Make a design proposal by selecting from their ideas and giving reasons for their choices.**	
Use a variety of gardening tools to prepare the soil. Describe the different uses of gardening tools and use them with care.	**AT 3 Level 1 & 2**	*explore and use variety of materials to design and make things.* *recognise that many materials are appropriate for different tasks.* *use materials and equipment safely.*	**Use a variety of materials and equipment to make simple things.** **Describe to others how they are going about their work.** **Show they can use simple hand tools, materials and components.**	**SCIENCE** Be able to describe simple properties eg colour, shape texture. Label the parts of a plant eg roots, leaves. Know that there is a variety of living things. Know that plants need certain conditions to sustain life. Group objects according to their characteristics.
The children were encouraged to judge their own efforts in producing healthy vegetables. Describe to the group how they planted, watered and harvested the vegetables and what they would do next time.	**AT 4 Level 1 & 2**	*evaluate their finished work against the original intention.*	**Describe to others what they have done and how well they have done it.** **Describe what they like / dislike with their designs.**	**ENGLISH** Participate as speakers and listeners in a group. Describe an event Communicate with the teacher, listen, ask and answer questions. Respond appropriately to simple instructions.

Figure 14

Figure 15

Melland School Sample Group Assessment Sheet **SUBJECT / AREA** Information Technology

AT	LEVEL	ACTIVITY	PROGRAMME OF STUDY Pupils should be taught:	WORKING WITHIN SoA Pupils should be able to:	OBSERVATIONS
5	1b	CHRIS : was assisted to turn the knob on the cooker to correct the temperature	that control is integral in many everyday products, eg cookers	talk about ways in which domestic equipment responds to signals or commands	Chris needs extensive guided instructions when working with the cooker. More emphasis needs to be placed on safety aspects.
5	1a	ELIZABETH : used the computer to type invitations to parents to invite them to come to the 'Vegetable Soup Party'	that information can be used to help plan and organise ideas in written form.	work with a computer.	Elizabeth can access the 'menu' using appropriate keys. She is learning to operate the space bar. With assistance she can obtain a print out of her own work.
5	2a	JAMES : worked alongside the teacher, using a graphics programme to draw a plan of the garden	to know that information can be held in a variety of forms.	use computer-generated pictures, symbols, words and phrases to communicate meaning.	James can identify the computer, monitor, printer and floppy disks. Using turtle graphics, he assisted the teacher to draw up a plan of the 'Vegetable Garden'.

ADDITIONAL IDEAS:

- Using the computer to record the most commonly liked vegetables.
- Use a variety of drawing programmes with which to sketch ideas for design, eg - for a herb garden.
- Create a vegetable word / picture matching game using the concept keyboard.
- Create own story book based on, eg 'The Hungry Caterpillar', using word processor and concept key board.
- Use microwaves to cook or heat up various foods.

the vegetables, displayed on the table nearby. Elizabeth picked up a potato and held it up to Noreen's nose. James, who had previously made his own book of vegetable likes and dislikes, walked over to the class library, found his book and showed it to the rest of the group.

Teacher: "What vegetable pictures should we put in OUR book?"

Elizabeth: "Carrots, like carrots."

James: "Onions." He made a sad face, "I cry."

Shazad picked up a large cabbage. (>>> AT2 Generating a Design)

Carl pointed to a turnip.

Elizabeth thought that Peter would like onions.

Noreen pointed to the lettuce.

The teacher then asked the children to think about how they would make the book and what materials they might need (AT1). It was decided to use paper, crayons, scissors and felt tip pens (AT3). The children chose their own sheet of paper and divided into two groups, to illustrate their chosen vegetables. Elizabeth helped Noreen to choose and use the pieces of vegetables to make potato prints. Don held Shazad's hand and helped him draw a green picture, pointing out that it was the same colour as the cabbage. James asked if he could use the computer and printed his choice of vegetable, copying the word from a label on the wall frieze (AT3). When everyone was finished, the teacher asked the children to choose the pictures they would like to choose for inclusion in their recipe book, for there was some duplication in the vegetables that the children had drawn. The children were encouraged to comment on one another's pictures and make suggestions (AT4).

'Potato, Bean and Onion' by James

Teacher: How can we put the pages together?"

Peter: "Glue it." (>>>AT3 Planning and Making)

Elizabeth picked up a paper clip "Look, this good."

James: "Get staple, stick together." (remembering how he had made his previous book)

The children decided that James's idea was the best and Carl and Elizabeth offered to staple the pages together, supervised by James who showed them where to put the staples. There was lots of discussion about the sequence of pictures in the book, the teacher encouraging the children to work out the order of the pictures, on their own. Eventually it was finished.

Teacher: "We've made our recipe book, now. Do you like it?"

James: "What's that?" (pointing to Shazad's interpretation of a cabbage) (>>> AT4 Evaluation)

Elizabeth included a picture of a cabbage, cut out from a magazine, to clarify Shazad's drawing. The teacher then turned the pages of the recipe book and pointed out that there was no cover page to let the reader see what sort of book he / she might be reading.

'Cover of the Recipe Book'

James: "I'll write it."
(>>> AT1 Identifying a Need
>>> AT4 Evaluation)

Elizabeth: "Put more pictures on it."

It was decided that a combination of both would be used. The children were asked to decide whether they wanted to keep the book in the class library or in the kitchen. They voted unanimously to keep it in the class library. This session was planned to promote open-ended, problem solving opportunities, thinking and discussion. Keith encouraged the children to try and solve the problems in different ways, keeping his own suggestions to a minimum. The tasks were structured so that each child could contribute and experience success throughout the session, which in turn motivated and sustained their interest and concentration. Figure 16 shows an individual Design and Technology recording sheet for James incorporating the objectives, set by the teacher and the subsequent assessment.

Sample Lesson 3 : 'Preparing & Making the Soup'

This lesson was a natural continuation of the recipe book made in the previous lesson. Since then the group had been involved in a series of related activities about vegetables which built on their previous experiences. They had visited a local supermarket to price vegetables and made a best buy chart. They had also spent time sorting the vegetables, according to the part of the plant they would eat, eg leaves from spinach, potatoes from bulbs and peas from seeds. This was illustrated on a bright colourful frieze on the wall behind the vegetable display table. The children had also written letters inviting parents to come to a free "vegetable soup party" but had decided to charge 10p a bowl for staff and children in school.

The teacher and the children had harvested a large number of vegetables from their vegetable garden and different groups of vegetables had been sorted into baskets and laid out on a table nearby. On a separate table was a selection of utensils, needed to prepare the vegetables and two large bowls of water to wash the vegetables.The children were again working in groups. Peter said he hated dirty carrots and the teacher initiated a group discussion about how they were going to do to clean the vegetables. Noreen and Greg, helped by the nursery nurse began to wash the vegetables and Elizabeth and Peter volunteered to dry them. Noreen giggled as she dipped her hands in and out of the water whilst Greg was absorbed by the cabbage leaves sinking and floating as he tried to press them down to the bottom of the bowl.

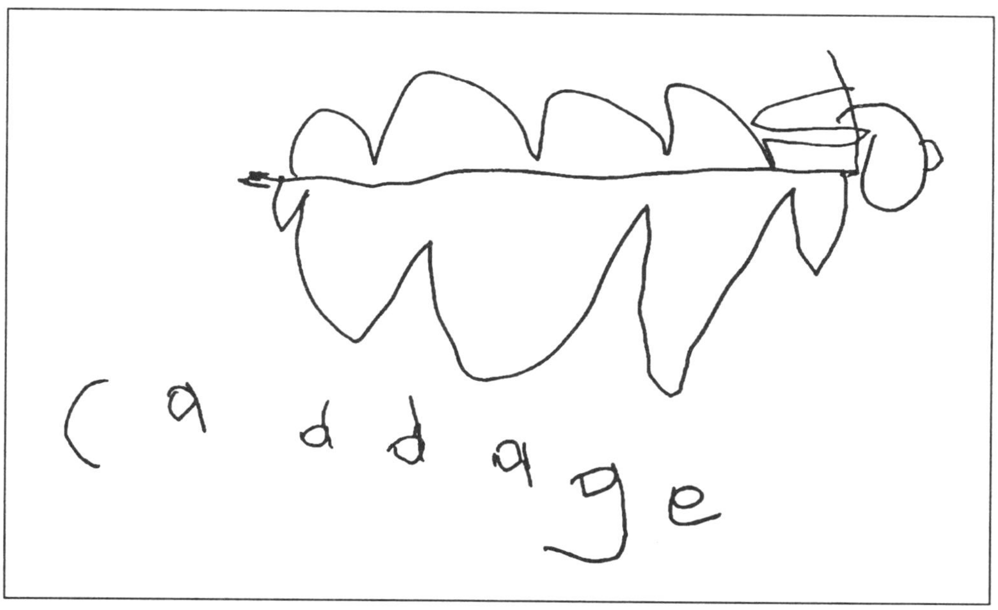

'Cabbage' by Elizabeth

When the vegetables were ready, the teacher gathered the group together and asked them to look carefully at the different vegetables, comparing size, shape, colour and weight. Elizabeth noticed that the carrots still had green leaves on top and the turnips were full of roots. There was considerable discussion about the viability of eating the leaves and roots. James suddenly realised that there was a frieze on the wall to identify which parts of the plant they could eat and pointed it out to the group.The children touched and smelt the freshly cleaned vegetables and compared them to the unwashed vegetables commenting on the different colour and texture.

Figure 16

Melland School **Individual Scheme of Work** **SUBJECT / AREA** Design & Technology **NAME** James

AT	LEVEL	DESIGN & TECHNOLOGY Statement of Attainment	DATE SET	TARGET BEHAVIOUR	EVALUATION
1	1a	Describe to others what he has noticed in familiar surroundings	June 14 1991	James will name and count the following vegetables - carrots, onions, potatoes, cabbage and turnips.	James identified and named potatoes, carrots and onions. Needs further work on discriminating between cabbage and turnips. Will be encouraged to take part in vegetable matching and naming games and will make own vegetable book.
2	2a	Use talk, pictures drawings models to develop his design proposals	June 21	James will make his own model vegetables using salt dough, paying attention to shape, size and colour.	James enjoyed this activity and used real vegetables as a guide to creating his own salt dough vegetables. He was concious of shape and colour but needed assistance to create relative sizes of individual vegetables.
3	1a	Use a variety of materials and equipment to make simple things	June 28	James will take part in the whole group activity of making a recipe book. He will select card, drawing implements and use scissors with care. He will help staple the pages of the book together.	James concentrated on drawing his vegetable pictures for a very short time. He needed several verbal prompts to stay on task. He used the scissors and the stapler with some regard for accuracy and quality.
4	1b	Describe to others what he likes and dislikes about familiar artefacts, systems and environments	July 4	James will be asked to identify his own vegetable preferences and explain why.	James named potatoes and onions as his favourite vegetables. He stated that he hated the taste of cabbage because "it was horrible." He informed the group that he liked the colour of carrots.

The teacher asked the children to predict the number of vegetables they would require for the soup and how to prepare them for cooking. The children needed considerable guidance for this activity.

Teacher: "How many carrots do you think we need?"

Elizabeth: "Lots and lots..." (>>> AT1 Identifying a Need)

Peter (who dislikes carrots): "Only three."

James decided that the class recipe book would provide the answer, so he brought it over to the group who counted the number of vegetables they needed.

Teacher: "What should we use to prepare the vegetables?"

James: "Need a potato peeler."
(>>>AT2 Generating a Design, >>> AT3 Planning & Making)

Elizabeth: "Need a knife" (to cut up the carrots).

Peter: "My mum pulls the leaves" (cabbage).

Throughout this discussion, the children were encouraged to listen to others, offer their own suggestions and explain the reasons for their choices. Greg distributed aprons to each child, Elizabeth and Peter chose to peel and cut carrots. Shazad, helped by Don, pulled the cabbage leaves apart, James chopped up the turnip and finally they all took it in turns to peel and slice the onions.

The teacher used this opportunity to question their actions, eg "James, are you going to slice the onions into small or big pieces ? (AT 2 & 3) Why?" ... "Elizabeth, is that the safest way to hold the knife?" (AT 4) The children were also encouraged to explore the use of different tools to prepare their vegetables and explain to the others the reasons for their final choice. Figure 17 illustrates how the individual needs of the children are being met through the selection of differentiated and age-appropriate activities.

Periodically, the teacher asked the children to describe what they were doing and why. When the vegetables were prepared, the children were offered a variety of different sized saucepans and asked to select one that would be suitable for cooking the vegetable (AT 1). When Carl chose a small saucepan, the teacher encouraged him to try it out. Carl quickly realised that it was not suitable and eventually after trying various sized pans, found one that was big enough to contain all the vegetables (AT 4). James was then guided to switch on the cooker and Peter set the time (AT 5). The teacher asked them what they should use to stir the vegetables (AT 1). There was further discussion about the merits of using a metal and a wooden spoon and it was decided to try both. The children soon realised that the metal spoon became very hot, whilst the wooden spoon stayed cool. Finally, the soup was ready and the children scooped their own helpings into their bowls and tasted it.

The general consensus was that the soup looked "mushy," and tasted "good" (AT 4). The teacher encouraged the children to talk about what they liked or disliked about the different cooked vegetables and asked them to record their preferences on a large poster, using happy or sad faces, ticking or making a cross next to the vegetables - see Figure 18.

The children were then asked if they might choose different vegetables next time for their home made vegetable soup. Elizabeth thought she would choose peas, James

- corn and Peter reiterated that he hated onions and didn't want any more in his soup. This was the final session that was observed at Melland school.

Melland School SAMPLE LESSON 3		
Individual & Group Objectives within an Integrated Scheme of Work 'Making vegetable soup'		
James	will type invitations to parents, to come to a Vegetable Soup Party. **Maths, Information Technology, English**	**Use IT to communicate meaning. AT5 Level 2a**
Noreen	will hold a soft nylon brush and be assisted to wash the potatoes. **Self-help**	**Use hand tools appropriate for the task. AT3 Level 2c**
Carl	will distribute aprons to the children in the group before preparing the vegetables. **English, PSE**	**Identify ways in which the work should be organised. AT1 Level 1a**
Shazad	will choose one vegetable and hold it for 5 seconds. **English**	**Work co- operatively. AT2 Level 2a**
Elizabeth	• will use a sharp knife safely to cut up the vegetables. • will describe her actions to the group. **PSE, English, Science**	**Use a variety of materials and equipment safely. AT3 Level 1a**
Greg	will feel a variety of vegetables to enhance tactile awareness. Observe reactions and monitor likes and dislikes. **Sensory and Science**	**Describe to others what they like about individual vegetables. AT3 Level 2a**
Peter	will turn on the cooker and set the timer. **Maths, Science, Information Technology**	**Recognise that the heat can be controlled by turning the knobs on the cooker. AT5 Level 1b**
Whole group	• will work co-operatively to prepare the vegetables. • will observe and stir vegetables in the saucepan. • will wash their hands before and after the cookery session. **PSE, Self- help, Science**	**• Talk about what they are doing during soup making. AT 1 Level 2a** **• Use a variety of equipment. AT 3 Level 1a** **• Evaluate taste, texture and colour of soup. AT 4 Level 1b**

Figure 17

PUPIL EVALUATION SHEET FOR VEGETABLES		
Draw / paste a picture next to the vegetable word to assist children who cannot read		
VEGETABLE	***I like it ...***	***I don't like it ...***
Turnip		
Potato		
Carrot		
Onion		
Cabbage		

Figure 18

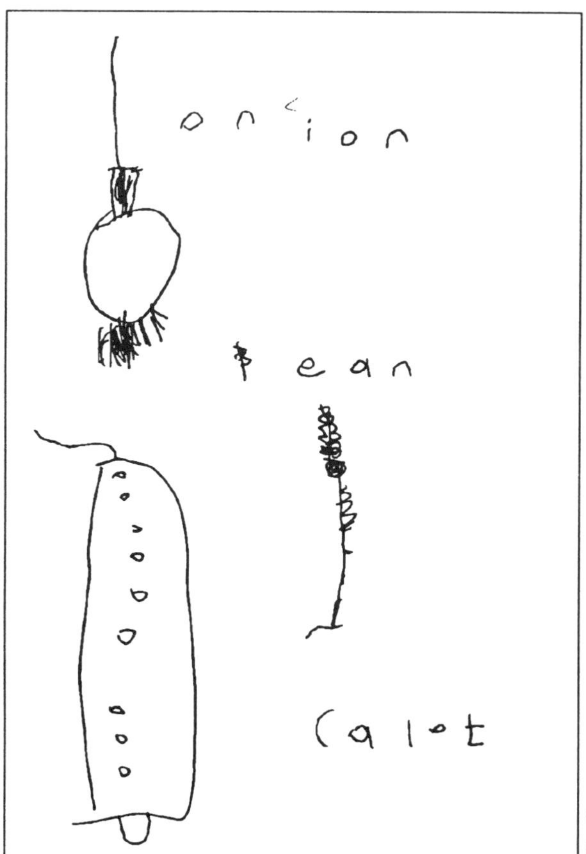

'An Onion, Bean and Carrot' by Elizabeth

Finally Figure 19 illustrates the ease with which the activities within the lesson "Making Vegetable Soup" incorporates the five attainment targets and ensures that the programmes of study are addressed. It also demonstrates how the core (cross-curricular) skills, themes and dimensions contribute to the enrichment and diversification of both, whilst providing ideas for new, relevant activities.

Sample Lesson 3	**Cross-Curricular Links : 'MAKING VEGETABLE SOUP'**		
ACTIVITY	**D & T AT'S**	**DESIGN & TECHNOLOGY PoS Pupils should be taught to:**	**CORE SKILLS, THEMES & DIMENSIONS**
Brainstorm ideas about ingredients for the soup.	**AT 1 Level 1 & 2**	*identify what should be done and ways in which the work is organised.* *use imagination and their own experience to generate and explore ideas.*	**SKILLS** • **Numeracy - Estimating the number of vegetables needed for soup.** • **Communication - Listening and giving weight to the opinion of others.** • **Study - Focussing on the task at hand.**
Describe to others what vegetables they can see and might use for vegetable soup.	**AT 2 Level 1 & 2**	*represent and develop ideas by drawings, models, talking, writing, working with materials.*	• **Problem solving - Making choices and cooperating with one another to solve a problem.** • **PSE - Washing hands before cookery.**
Discuss different ways of preparing and cooking the vegetables.		*explore and use a variety of materials to design and make things.*	• **IT - Supports learning across a variety of curriculum areas.**
Draw pictures of the vegetables they like and dislike.		*recognise that many materials are appropriate for different tasks.*	**THEMES** • **Economic and Industrial Understanding - Writing out ingredients and recipe for a meal.** • **Health Education - Practice simple personal routines, eg washing hands.**
Use a variety of utensils to prepare the vegetables.	**AT 3 Level 1 & 2**	*use materials and equipment safely.*	• **Careers Education and Guidance - Plan to visit to a gardener to discover nature of job.**
Describe their actions to the group members.			• **Environmental Education - Using their senses to discover**
Discuss how to use the cooker.	**AT 5 Level 1**	*that control is integral in many everyday products eg a cooker.*	**the natural world, eg identifying the vegetables by smelling the leaves.**
Turn on the timer.			• **Education for Citizenship- Group visits to garden centres and nurseries.**
The children were asked to comment on the vegetable soup, its appearance texture and taste.	**AT 4 Level 1 & 2**	*Pupils should be taught to: evaluate their finished work against the original intention.*	**DIMENSIONS** • **Equal opportunities - Boys and girls were encouraged to take part in harvesting and cooking.** • **Differentiated activities to allow all pupils to take part.**
They were encouraged to make suggestions for improving the taste, eg - adding different vegetables.			• **Allowing the pupils real choices and the opportunity to make decisions and mistakes for themselves.**

Figure 19

Case Study 3 - Upper School: 'The Biscuit Factory' (Leacroft School)

Leacroft School is a sixty place, group 1s establishment for pupils, aged two to nineteen, with severe and profound learning difficulties. All pupils with profound and multiple learning difficulties are integrated into classes with their peers.

The Senior department is comprised of sixteen students divided into two classes. Under the mantle of TVEI, it has begun to adopt a cross-curricular, modular approach to the acquisition of life skills and attitudes, within a real life context. Through work experience and mini-enterprise projects, the students are involved in activities that promote self-advocacy, extend independence at school and in the community and develop their inter-personal skills. This case study examines one module which was carried out in the spring term at Leacroft School.

The objective for the module, 'The Biscuit Factory', arose out of the need to promote extended integration links with Manchester College of Arts and Technology, Openshaw Centre. This project also fulfilled a secondary aim which was to generate extra money for a planned residential holiday for the students, during the summer term .

Figure 20 sets out the student objectives for the module within TVEI and Figure 21 highlights the technological process that took place within "The Biscuit Factory." At the outset the students were involved in planning the the long term objectives for the module.

Sample Lesson 1 : Planning 'The Biscuit Factory'

The students and Mike (the teacher) sat in a semi-circle discussing and sharing ideas for the creation of The Biscuit Factory which would involve making and selling biscuits - in effect, running a small business. During the previous week, they had visited Sainsburys on two separate occasions to observe and record information on the best selling biscuits. They asked some shoppers to name their favourite biscuits and observed other customers, to see which ones they chose. The students had then recorded the results and learned that oatmeal biscuits were the best selling, most popular biscuits. The students had also taken a selection of commercially produced biscuits into Openshaw centre at break time and distributed them free of charge. Openshaw students also preferred the oatmeal biscuits.

A TVEI Enhanced Course

School / College *Leacroft*
Course Title *The Biscuit Factory*
Enhanced by
Lessons per part *2 per week*

Main Contact *Mike Hall*
Date trialled *Spring 1991*

Objectives for the module

To provide students with the opportunity to integrate with studentsfrom Openshaw College.
To experience the planning and commitment needed to set up a small business.
To make choices and decisions, within a group.
To utilise skills in communication, functional number / money.

Methodology

The objectives outlined above will involve working in a group under the guidance of a Leacroft teacher.
The students will decide the product, name of biscuit, name of logo for T-shirts.
The nature of each session will be dictated by the pupils themselves.
A number of activities will be covered in this module: Marketing, Cookery, Health and Safety and Book-keeping.

Resources

BBC Computer
School House (Kitchen)
Ingredients for biscuits
Paper, card, pens, scissors and glue to attach the label to the wrapped biscuits.
Manchester Polytechnic School of Education Art Department to use the printing facilities and equipment for printing own T-shirts.

Methods of Assessment

The student's experience / learning / knowledge will be assessed, using teacher observation supplemented by video recordings and photographs.
Self-assessment by the students of the success of The Biscuit Factory.

Equal Opportunities: Outcomes

Students will take their place in the community.
Integration with students and staff at Openshaw will reflect this.

TVEI Planning Sheet - 'The Biscuit Factory'

Figure 20

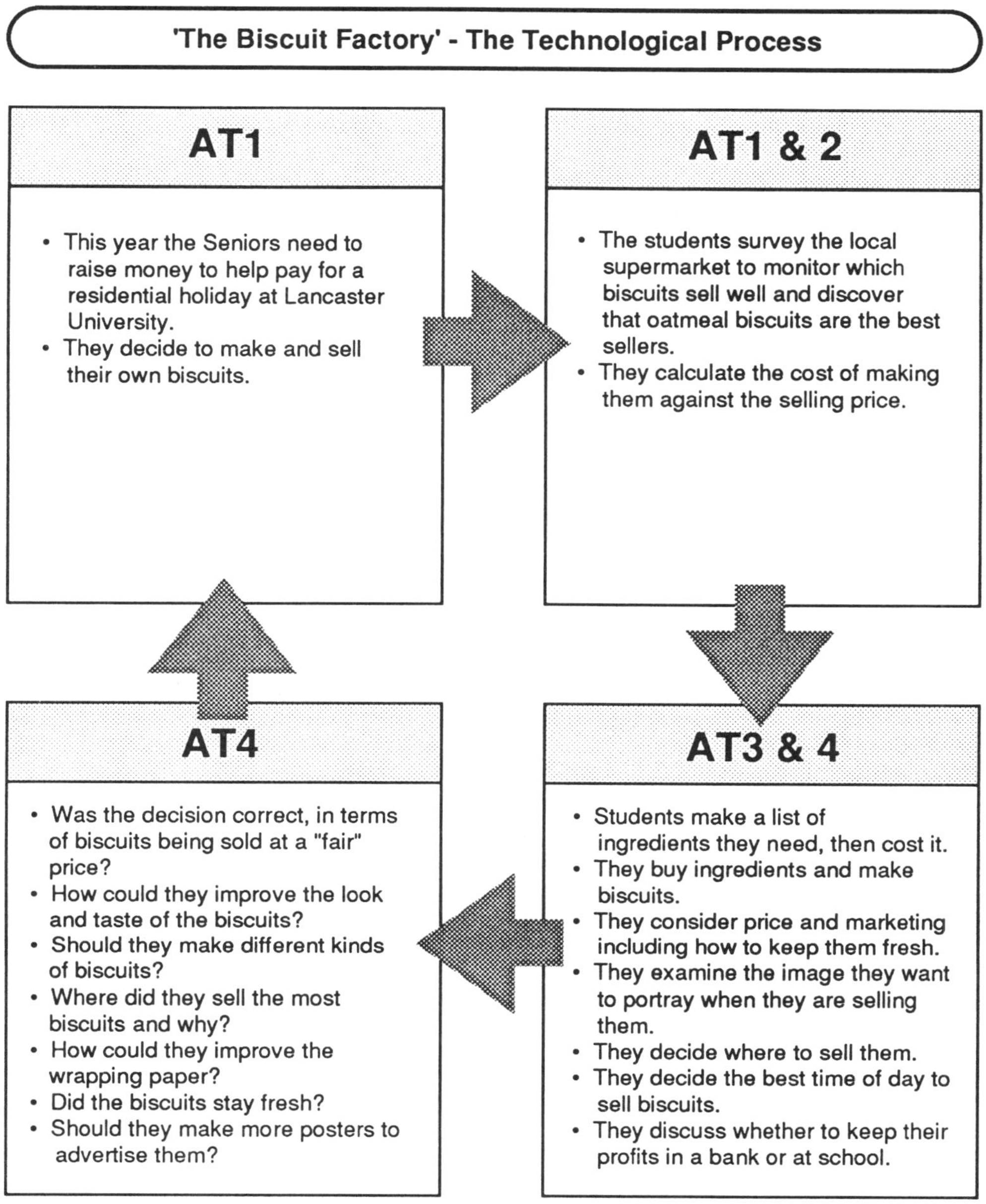

Figure 21

The teacher reminded them about their forthcoming holiday and the money making purpose of their biscuit making project. He asked the students to identify a variety of biscuits, arranged on plates on a nearby table. Janet, a student with PMLD was offered a selection of biscuits (which were crumbled up) to taste and her reactions were observed. Marion sat next to her, acting as her advocate. These two students have very different needs. Figure 22 illustrates how the individual educational objectives of both students were being met within Sample Lesson 1.

The students decided that they would make oatmeal biscuits and there was considerable discussion about whether to call them a special name to encourage shoppers to buy them. David, who had just been reading "Not now, Bernard" began to talk about Monsters and Marion laughed, saying" MONSTER COOKIE."

All the students agreed that it was the best name for the biscuits. One of the students had brought a simple recipe from home and Marion (an independent reader) read the

Figure 22

Leacroft School Differentiated Technological Activities Within Sample Lesson 1

Priority Targets	AT1 Identifying Needs & Opportunities	AT2 Generating a Design	AT3 Planning & Making	AT4 Evaluating
JANET 1. Encouraged to: • show differential responses eg - *vocalising, stilling, reaching* • to indicate "I like" • *head turning, thrashing* • to indicate "I don't like" 2. Can locate sound source. 3. Janet will maintain brief eye contact 4. Encouraged to explore objects, different smells, tastes, surfaces and materials.	Janet opened her mouth in readiness for tasting. Janet relaxed & made hand / finger reaching movements towards the biscuits.	Janet was helped to press a switch on the computer to assist Marion to print the results of the survey.	During the visit to Sainsburys, she touched the wrappings of several different packets of biscuits. Janet gave brief eye contact to Marion.	Janet laughed when Michael made Monster noises. After eating one piece of biscuit, Janet was observed reaching for another.
MARION 1. Marion will speak clearly when she "interviews" strangers. 2. Will read stories to the group. 3. Will use the computer and the word processor to type and print out own work. 4. Will use own initiative in unfamiliar situations.	In Sainsburys Marion initiated and maintained a brief conversation with a customer. Marion shared her ideas with the group Listened without interrupting to others.	Suggested making very big cookies. Drew a picture of her favourite biscuit. Recorded the results of the survey. Noted the preferences of others.	Used crayons and felt tips to draw her favourite biscuit. Used a recipe book to read the list of ingredients that would be needed for making "Monster Cookies".	was too expensive for one biscuit. Noted that Janet stiffened to show dislike. Aware of the difference in price between a small and big biscuit.

list of ingredients to the other students. The students shopped for the ingredients and then costed them. Mike asked them how much to charge for each biscuit:

David:	"£1." (>>> AT1 Identifying a need)
Tracey:	"£50."
Marion:	"Don't be daft."

'The Cookie Monster says "Buy My Biscuits"', by David

Mike showed them a large and a small biscuit and asked them if they should charge the same for both. Marion looked uncertain and the rest of the group didn't know. The students required considerable guidance to decide that the small biscuits would cost less than the big ones. Eventually, the price was set at 5p for a small biscuit and 10p for a big biscuit. Throughout the session, the teacher acted as group facilitator, encouraging the students to make choices, solve problems and communicate their own ideas with other members of the group.

During the following week the students made a batch of 30 biscuits, which involved many cross-curricular skills outlined below :

- Estimating: the amount of flour, oatmeal etc they would need
- Weighing: different ingredients
- Measuring: amounts of liquids
- Observing: changes during cooking
- Communicating: with each other, sharing their observations
- Decision making: what kind of biscuits to make.

As these skills have been covered in some depth within the cookery lesson (Vegetable Soup) outlined at Melland School, it is the purpose of the next sample lesson to examine the preparation of the biscuits to be sold at Openshaw Centre.

Sample Lesson 2 : 'Wrapping the biscuits & creating a logo'

During the week, Mike had invited a Health and Safety officer to talk to the class about personal and food hygiene when handling food. The students were then divided into two groups to solve the following problems:

- Finding the best way of wrapping the biscuits, to keep them fresh. (>>> AT 2 Generating a design)
- Creating a label that encouraged people to buy them.
- Designing and making a logo for the biscuit posters (whole group activity).

The students were divided into two mixed ability groups:

Group A

Marion, Janet, Michael and Tracey, working with Diane (nursery nurse) were exploring different types of materials, to discover which one offered the most protection for the biscuits. A selection of tissue paper, cardboard, plastic sheeting, aluminium foil and cling film was arranged on a table. The students were encouraged to test the paper for, eg ease of wrapping, whether the material was waterproof and the appearance of the wrapped biscuit. Janet was encouraged to feel a selection of paper and her reactions were observed and noted. She appeared to enjoy the smooth, slippery feel of the cling film and plastic sheeting but quickly removed her hand from the tissue paper. When Tracey tried to wrap up a biscuit in tissue paper, it began to tear and she soon discarded it in favour of the aluminium foil. However, that also ripped so she experimented with the cling film. She found it difficult to cut and asked Marion to hold it while she estimated the amount she needed to cover the biscuit. Michael could not decide which material to use so Diane (nursery nurse) encouraged him to choose one, test it out and decide if it was suitable. Marion immediately chose the cling film because, "I can see the biscuit inside." The students decided that the biscuits wrapped in cling film were best because " they looked shiny."

Group B

Kelly, David, Lyndsay and Mike were using a variety of materials, exploring different ways of labelling the biscuits. Felt tips, crayons, paint, pens and the computer printout were all tested. Mike had asked them to find a quick way to label them. David tried to decorate the wrapped biscuit with poster paint, splattered paint on Kelly's jumper and accidentally smeared some on his own face. He observed that paint made the cling film very soggy. Lyndsay and Kelly were drawing pictures, cutting them out and pasting them on the biscuits. This took a long time and after twenty minutes they had only labelled one biscuit each. There was considerable discussion, therefore about having to repeat this activity thirty times each week (as mentioned earlier, the students had decided to make 30 biscuits each week).

The teacher asked them if they could think of a way of using the computer to make the labels (AT 2). Kelly liked this idea and using a written model of Monster Cookie, typed, printed and glued it onto the wrapped biscuit (AT 3). The students agreed that it had only taken a couple of minutes to do this and Mike pointed out that if they made 30 photo copies of one label, then they would have enough for each batch of biscuit (AT4). Throughout the lesson the students were continually encouraged to make choices about when and how to use IT in their work (AT 5). Figure 23 offers a sample of IT activities that the students used to support their Design and Technological activities during Sample Lesson 2.

The students were then asked to think about a logo for the biscuit posters to be distributed at Leacroft School and Openshaw Centre. Michael suggested "A big monster," and the students liked that suggestion.

'One of the Dragons'

All the students created their own monster picture and then decided which one they liked the best. Marion's picture of a huge green, blue and yellow dragon received the most votes. The final activity was more difficult for the students. They were encouraged to make up a slogan to print on the poster. After much deliberation and adult intervention, they were asked to choose one of the following:

- EAT THEM BEFORE THEY EAT YOU (Mike's suggestion, the eventual winner)
- DRAGONS EAT BISCUITS (Marion's suggestion)
- MONSTERS ARE SCARY (Michael's suggestion)

Marion and Michael shared the job of printing the winning slogan on the poster and Kelly and David used the photo copier to make ten copies. The lesson ended with a group discussion about all the work they had completed that morning and acknowledgement of each other's considerable accomplishments. Figure 24 summarises the preceding activities, together with the relevant Design and Technology programmes of study and attainment targets covered within this session.

Figure 23

Leacroft School **Sample Group Assessment Sheet** **SUBJECT / AREA** Information Technology

AT	LEVEL	ACTIVITY	PROGRAMME OF STUDY Pupils should be taught:	WORKING WITHIN SoA Pupils should be able to:	OBSERVATIONS
5	2b	MARION : entered the results of the 'Biscuit Survey' into the computer.	how to store, select and analyse information using software, eg simple data base package.	Use information technology for the storage and retrieval of information.	Marion is showing increased confidence when using the computer. She is using the computer on a regular basis to reinforce work in other areas.
5	1b	DAVID : used the photocopier to print ten copies of the slogan "Eat them before they eat you."	how to give instructions to electronic devices.	talk about ways in which equipment responds to signals and commands.	David needed assistance to instruct the photocopier to print 10 copies. He kept pressing the 0 first then the 1. Finally, he used a model and guided instruction to work from left to right.
5	2a	KELLY : used a written model of "MONSTER COOKIE" to type and print out label for the wrapped biscuits.	to know that information can be held in a variety of forms.	use computer - generated pictures, symbols, words and phrases to communicate meaning.	Kelly could locate the following letters on the keyboard - O, E, S and T, but found it difficult to distinguish M from N. She is learning to access the software using a 'mouse'.

ADDITIONAL IDEAS:

- Use of the computer and word processor to create own biscuit recipe book, recording investigational work.
- Use of video and camera to record 'selling biscuits' at Openshaw College. Students are given the opportunity to set up the TV, insert the video, rewind and play.
- Use of bank card to withdraw money from the automatic bank dispenser machine.
- Tape recording 'interviews' with students at Openshaw College to determine whether MONSTER COOKIES are eg: good value for money.
- Use the phone to make contact with the College to make arrangements for the next visit.

Session 2	Design & Technology : 'THE BISCUIT FACTORY'	
ACTIVITY	**D & T AT's**	**DESIGN & TECHNOLOGY PoS**
Discussion about the best material for wrapping the biscuits. Drawing own posters. Typing and printing the labels for the biscuit.	AT 1 Level 1	<u>Developing and communicating ideas</u> *Pupils should be taught to:* *• use imagination and their own experiences, to generate and explore ideas.* *• represent and develop ideas by drawings, models, talking, writing, working with materials.*
Discussing the choice of material for wrapping the Monster Cookie. Sharing ideas about how they might wrap and label the cookies, making reference to the criteria set, eg - is it waterproof? - can you read the label? - does it look attractive?	AT 1 Levels 1 & 2 AT 2 Levels 1 & 2	<u>Developing and using artefacts, systems and environments</u> *Pupils should be taught to:* *• know that a system is made of related parts which are combined for a purpose.* *• identify the jobs done by parts of the system.* *• give a sequence of instructions to produce a desired result.* *• recognise and make models of simple structures around them.* *• identify what should be done and the ways in which work should be organised.*
Designing and making a poster. Exploring different ways of labelling the cookies, using paint crayons and felt tip pens.	AT 3 Level 1 & 2	<u>Working with materials</u> *Pupils should be taught to:* *• explore and use a variety of materials to design and make things.* *• recognise that many materials are available and have different characteristics which make them appropriate for different tasks.* *• use materials and equipment safely.*
Evaluating their own poster and those of others in the group.	AT 4 Levels 1 & 2	<u>Satisfying needs and addressing opportunities</u> *Pupils should be taught to:* *• evaluate their finished work against the original intention.*
Using a computer and printer to create own label. Using the photocopying machine to duplicate pictures.	AT 5 Levels 1 & 2	*Pupils should be taught that information technology can be used to help plan and organise ideas in written and graphical form.* *Organise and present ideas by using a simple word processor.*

Figure 24

Sample Lesson 3 : 'Selling the Biscuits'

This session was a continuation of activities begun in the previous lesson. The students had wrapped and labelled a batch of Monster Cookies and prepared several biscuit posters to advertise at Openshaw Centre.

'Cookie Monster' by Marion

The teacher asked them to consider how they would carry thirty biscuits to the Centre and to decide how much money to take with them, to provide the necessary change:

Mike "How are we going to carry the biscuits?" (>>> AT 1 Identifying a need)

David "In my pocket."

Tracey "In a big bag."

Marion (laughing) "Put them in a wheelbarrow."

It was decided to carry the Monster Cookies to Openshaw in a tin box, so they would lie flat. Tracey offered to look after them on the mini bus, helped by David, who was worried that she might eat some on the way.

Mike had provided a range of bags, boxes and an old ice-cream tray, which were displayed on a nearby table and asked the students to pretend they were serving customers and decide which container would be the best for that purpose. Marion strapped the ice-cream tray round her shoulders and filled it up with biscuits. She complained that it was too heavy and did not want to use it. Michael suggested using a paper bag but that was awkward to manage. After trying a number of containers, it was decided that the students would try setting out a table to display the Monster cookies. This seemed to work well and the group agreed to use this method. Marion had found a small plastic sorting tray, in which to keep the change and profits, which Janet would hold on her lap.

One of the priority targets for Marion and Michael, set down by Mike at the beginning of the module, was the opportunity for both to gain further experience of travelling independently to known destinations. Therefore, during the previous week, he had spent some time with them both, reviewing the route that they would take to Openshaw Centre, the cost of the bus fare and what they would do if they were lost. He had arranged for the two students to be "shadowed" by a volunteer, for this was only the third time they had travelled to Openshaw by themselves. The remaining students travelled by minibus to the Centre.

Once everyone arrived at the Centre, the students chose where to place the posters advertising the Monster cookies. Mike encouraged them to think about putting them where they would easily be seen. The selling of the biscuits was a success. All the students took part and at the end of break time, there were only four biscuits left. There was lots of friendly interaction between the two sets of students, which culminated in an invitation to the Leacroft students to return to Openshaw Centre the following week.

When the Leacroft students arrived back at school, they discussed the visit and Mike encouraged them to evaluate the experience in the light of the following observations:

- Some of the Openshaw students complained that some of the cookies were too thin.
- Some students unwrapped the cookies, then dropped the cling film on the floor, creating litter.
- Some of the wrapped cookies fell out of the wrapping paper, when they were picked up.

As a result of these observations, several extended activities were devised and are outlined, below:

EXTENDED ACTIVITIES

- The students visited United Biscuits, a factory based in Stockport, near Manchester to observe the manufacture and packaging of biscuits.

- They recognised the need to improve their own wrapping paper and experimented with cutting different sizes and shapes of cling film to ensure that the biscuits were totally protected.

- There was considerable discussion about what they would do if a customer asked for more than one biscuit - the need to create a "package."

- The students explored using different sized cutters to create a uniform biscuit - consistent thickness, size and shape.

- They also discussed the need for a waste paper basket that would accommodate the discarded cling film.

However, all the students agreed that they would like to go to Openshaw Centre again. Marion told the group that she was going to wear her own special purple T-shirt next time.

This sample lesson illustrates clearly that the four phases of Technology are not necessarily sequential, but occur throughout the process at different times, one often generating the other. For example, evaluation of David's painted label for the Monster Cookie, provided the stimulus to identify a need to use and therefore design something different. Figure 25 summarises the differentiated activities and cross-curricular implications of the sample lesson, "Selling Biscuits".

Sample Lesson 4 : 'Printing the T-shirts'

The Biscuit Factory was underway. It was cross-curricular and experiential learning at its best. The students had become market researchers, sales assistants, buyers, advertising agents and book - keepers, the essential personnel for running a business. At this point, Mike introduced the concept of the image that the students wanted to portray, when they were selling their Monster Cookies. The following issue needed to be addressed:

- How would the students dress when selling their Monster Cookies ?

Marion's comment about wearing a special T- shirt prompted more discussion, and it was decided that the students would print their own special T- shirts, that would be worn when selling "cookies" at Openshaw Centre. Therefore arrangements were made to visit a graphics design centre within Manchester Polytechnic. The students chose to use the logo of Marion's dragon with the accompanying slogan "Eat them before they eat you." (See Sample Lesson 2)

The following week, the students arrived at Manchester Polytechnic, ready to print their T-shirts. John, the instructor explained the procedure to them. The students were encouraged to work in pairs, so they could help one another prepare the boards which needed to be covered with newspaper and the T- shirts. Michael and David worked together to cover one another's boards and negotiated with each other who was going to hold the sellotape and who was going to cut it.

The next step in the process was preparing the colour. John explained that the outside colour, blue would be printed first and there was considerable discussion about the shade of the colour. Was it a dark or light blue? How do you make a colour darker or lighter? The procedure for mixing the colour was demonstrated and Marion volunteered to do it by herself, spooning the white base paint into a bowl. Michael thought it was toothpaste, David said it was icing sugar and Tracey thought it looked like soft ice-cream. The blue colour was added, mixed into the white base to "make it smooth" and examined to see if it matched the shade of blue in the logo. Several students recognised that the shade was not the same and were asked to think of ways of making it lighter. Marion, who already had previous experience at mixing and matching colours, suggested adding more white paint to the mixture. Placing the colour on the frame of the screen, the instructor demonstrated how to hold the brush and pull it backwards and forwards across the screen. David found it difficult to exert enough pressure on the brush and asked Michael to help him. The screen was then lifted to reveal the first colour on the T-shirt and there was much excitement and interest in the result. Michael said he thought it was "magic". The procedure was then repeated until all the colours were printed on the T- shirts.

Throughout this session the students were totally engrossed in the activity, helping each other, offering advice, giving support and learning to share and wait until it was their turn for the tools that they needed. When all the T-shirts were completed, they commented on each other's work.

Leacroft School SAMPLE LESSON 3		
Individual & Group Objectives within an Integrated Scheme of Work 'THE BISCUIT FACTORY : Selling Biscuits'		
David	David identified a 5p biscuit and a 10p biscuit **Maths, English**	**TECHNOLOGY** **Knowing that goods are bought and sold.** **AT1 Level 2c**
Janet	Janet did not resist having the tray placed on her knee. **Communication, Self-help**	**Using materials appropriate to the task.** **AT3 Level 2c**
Tracey	Tracey explored different containers to carry the biscuits, looking at size and shape. **Maths, Science**	**Observing and describing familiar things.** **AT1 Level 2a**
Marion	Marion sorted the money into 5p, 10p and 20p coins in the cash box. **Maths**	**Identifying what should be done and ways in which the work should be organised.** **AT2 Level 2b**
Lyndsay	Showed pleasure and gave eye contact, when another student spoke to her. **PSE, English**	**Showing preferences.** **AT4 Level 1a**
Kelly	Kelly corrected a crooked poster that she had put on the wall. Initiated conversations with customers. **English, PSE**	**Evaluating things by observing them and suggesting practical changes.** **AT4 Level 1b**
Michael	Observed that the students at Openshaw Centre dropped the cling film on the floor. **Maths, Science, Information Technology**	**Describing to others what they have noticed in their surroundings.** **AT1 Level 1a**
The whole group	• Will work co-operatively to sell the biscuits. • Will integrate with students from Openshaw Centre. • Will take responsibility for displaying the biscuits on a table at Openshaw Centre. **English, Maths, PSE, Self-help, Science**	• **Expressing their ideas.** • **Describing to others how they are going about their work.** • **Making simple judgements.**

Figure 25

Sample of comments

Marion:	"Funky." (>>> AT 4 Evaluation)
Michael:	"Like mine best."
David:	"Like the yellow."
Tracey:	"Want to do a turtle, next."

At the end of the session John gave each of them a certificate award outlining their achievements.

Back at school, Mike and the students discussed the morning's activities. It had been a hard morning's work, but the students were the proud owners of their own printed T-shirts.

A variety of methods were used to record pupil progress across the different activities within "The Biscuit Factory." Video recordings, monitoring sheets, samples of student's drawings and writings were selected as evidence. Figure 26 is an illustration of an individual recording sheet for Tracey, whose activities were monitored over a period of a week within "The Biscuit Factory."

Two days later, the students, wearing their "funky" T-shirts returned to Openshaw Centre to sell another batch of biscuits. Many of the Openshaw students commented and admired the T-shirts, asking if they could order one from the Leacroft students. This was unexpected but welcome. Negotiations are currently under way to see if this is viable, for it has generated an idea and presented the students with an opportunity for further development. An outline of the benefits the students derived from running a school based business is indicated in the Table below.

DEVELOPMENT OF SKILLS THROUGH:		DEVELOPMENT OF ATTITUDES THROUGH:
Observing	visits to bakeries and shops seeing people as workers	teamwork, cooperation
Questioning	interviews, surveys	collaboration
Analysing	research & sales data	quality of presentation
Communicating	through discussion through advertising while selling & buying	relationships
Decision making	running the business	reliability, adaptability
Evaluating	as an on-going strategy	loyality & committment

Throughout this case study, the students were seen to be achieving the maximum benefit from the activities offered to them:

- The work was challenging.
- The tasks were structured so that each student could achieve success.
- The teacher's expectations were appropriate.
- The students' successes were recognised.

Figure 27 summarises the cross-curricular subject links between Technology and the other curriculum areas and highlights the cross- curricular elements that permeate all activities in working with pupils with severe and profound learning difficulties.

Leacroft School Individual Recording Sheet

MODULE MINI ENTERPRISE 'The Biscuit Factory'	**NAME** Tracey
AIM To take a shared responsibility in the setting up and running of a small business.	**CLASS** Senior 1
	TERM / YEAR Summer 1991
	KEY STAFF Diane, Mike

DATE	PUPIL IS WORKING TOWARDS	COMMENTS
June 10	• independently shopping for appropriate supplies for cookery • extending her numeracy skills through using money, including counting change	<u>Shopping for Supplies:</u> • Tracey needed some verbal prompting to find the necessary ingredients in the shop. • She paid at the desk, received the change and receipt and presented them to Mike. **Tech. AT1 Level 1**
June 11	• extending her ability to concentrate in the face of distractions • observing hygiene rules when working with food	<u>Biscuit Making:</u> • Tracey was responsible for cooking the butter / sugar / syrup mixture but she was distracted by Michael and let the mixture boil. **Tech. AT3 Level 2b** • She remembered to wash her hands before and after cookery. **Tech. AT2 Level 1a**
June 12	• persevering with a difficult task, asking for help when necessary • working co-operatively with her peers	<u>Packaging the Biscuits:</u> • Tracey struggled a bit with the cling film. She worked much better at sticking the labels on the wrapped biscuits. **Tech. AT3 Level 2c** • She worked efficiently with Michael. **Tech. AT2 Level 2a**
June 13	• maintaining extended conversations with a variety of people • giving accurate answers to questions	<u>Openshaw Centre</u> • Tracey worked hard selling biscuits. She was very pleasant and chatty with the Openshaw students. • She clearly indicated the different prices of the biscuits.

Figure 26

Figure 27

'The Biscuit Factory' & The Whole Curriculum

CROSS-CURRICULAR THEMES

Environmental Education	Health Education	Careers Education	Citizenship	Economic & Industrial Understanding
• Transport systems to & from Openshaw Centre • Awareness of litter & recycling	• Healthy eating • Food Hygiene • Personal hygiene	• Awareness of the implications of work • Visit to local Bakery	• Awareness of people working together to provide a service	• Advantages of buying ingrediants in bulk • Economics of a small business • Awareness of range of people in a work place

CORE SUBLECTS	COMMUNITY AWARENESS	PARENTAL INVOLVEMENT	CORE SUBLECTS
English	• Awareness of services provided in the community • Visits to Bakeries • Understanding of the number of jobs involved in running a small business	• Parental assistance with selling biscuits • Inviting parents to help make biscuits • Students practiced baking biscuits at home	**Design & Technology**
• Discussing ideas • Listening to others • Sharing ideas • Reading recipes • Labelling biscuits • Writing & recording information for the survey			• Designing wrapping paper • Making biscuits • Conducting Surveys • Planning, designing & making - T-shirts, Posters & Labels INFORMATION TECHNOLOGY • Using the computer for accounts & recording
Science	**CORE SKILLS**	**CROSS-CURRICULAR DIMENSIONS**	**Geography**
• Investigating changes during cooking • Tasting sessions • Smelling & feeling different substances • Healthy eating • Groupings of food • Origin of flour • Uses of syrup	• Making choices • Taking descisions • Communicating • Explaining • Listening • Solving problems • Negotiating • Evaluating • Organising • Cooperating	• Involvement of both male & female students • Whole class activity	• Ways in which biscuits are made in other countries • Origin of some of the ingredients, eg raisins
Maths			**History**
• Data collecting, eg - number of biscuits made per session • Simple book-keeping • Handling money • Estimating, weighing & measuring ingrediants			• Visits to museums to examine old cooking implements • Comparing old coinage before decimalisation
			Art
			• Mixing colours for printing on T-shirts • Drawing the labels for the wrapped biscuits • Creating a logo for the T-shirts

Adapted from Primary Schools & the World of Work Manchester City Council (1989)

Case Study 4 - Upper School: 'Materials & their Uses' (The Birches School)

The Birches is a Group 2s school for pupils with severe learning difficulties aged two to nineteen years and is situated in the inner city area of Manchester. There are seven class groups and each class has a teacher and up to two members of non-teaching staff, depending on the needs of the pupils. Pupils with profound and multiple learning difficulties are integrated into the leavers class with their peers.

The class group observed in this case study were aged fourteen to nineteen years and, like the group at Leacroft school, were working within the Technical and Vocational Educational Initiative (TVEI) modules. There were eight students in the group altogether, three of whom had profound and multiple learning difficulties. The school theme for the term was 'Materials and their Uses'. Figure 28 below illustrates a topic web which explores some of the ways in which the theme could be developed.

Figure 28

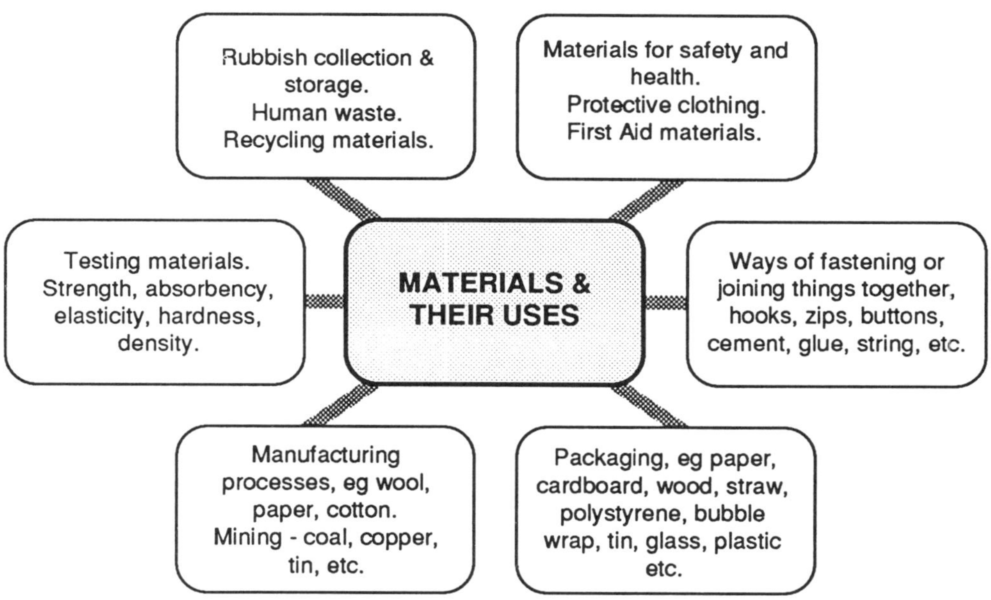

Suggested Activities for Termly Theme 'Materials & their Uses'

In the Spring term a link was set up between the class teacher at the Birches and two teachers from Manchester City Council's Inspection and Advisory Service who were based in a technology unit at the Abraham Moss Centre (a combined school, college and community centre in North Manchester). The advisory teachers were from mainstream schools and were working on Design and Technology activities with several schools throughout the city. In addition to working with pupils and students, the sessions were organised so that they also provided in-service training for members of staff at the Birches.

The Centre had much better facilities than the school, with a large workshop containing work benches, a wide variety of tools and materials. Because the sessions took place at an unfamiliar venue some distance from the school, it provided an additional valuable experience for many of the students who were able to develop some independence by finding their way around the building unaided.

Six lessons of one morning's duration were planned over a period of six weeks and the aims for the whole group were as follows:

- To introduce students to Design and Technology through the topic of materials.
- To introduce students to and give them experience of a wide range of materials emphasising the great variety of materials available.
- To introduce students to and give them experience of the variety of different uses of materials and to discover some of the different properties of materials.
- To introduce students to new tools and equipment as well as familiar ones.
- To develop safe handling of equipment and materials.
- To relate all tasks to National Curriculum attainment targets and students' individual learning programmes.
- To foster the development of new skills.
- To encourage creative skills and aesthetic awareness.

The staff from both establishments planned activities which could be generated from the students' own experiences. Some of these linked to work which was already going on in the school, but others were deliberately designed to allow the students to investigate their own ideas. The session on puppet making for example, developed as an extended activity from personal and social education lessons at the Birches. Students had been discussing the many issues surrounding talking to strangers and had been made aware that there were occasions when they might be in need of help. They explored the possibility that there might be some adults whom they could trust if they were lost, hurt or ill. The class teacher felt that some of the issues could be made more meaningful if students made puppets and created a statement about how that 'character' might help them. This presented opportunities for exploring many curriculum areas including the Design and Technology capability within personal and social education.

Planning continuity and progression within the module required careful consideration because the sessions at Abraham Moss Centre only took place on a once weekly basis. Follow-up work at the Birches also had to take into account the school's more restricted facilities. The activities observed at Abraham Moss Centre were:

- Designing and making puppets.
- Exploring structures.
- Exploring materials to make battery-powered vehicles.
- Paper sculpture (designing and making hats, flowers and masks).

Figure 29 provides a summary of the main activities which took place over the six week period at Abraham Moss Centre, together with the relevant PoS and AT's for Design and Technology and cross-curricular links within the National Curriculum.

For the purpose of this case study four of the six sessions will be outlined in detail to give an idea of the activities which took place at Abraham Moss Centre. The two sessions on designing and making vehicles will be discussed together. Observations will also address National Curriculum areas including cross-curricular links, the development of areas within the whole curriculum and the progression of activities at the Birches.

Session 1 : 'Puppet Making'

The workshop at Abraham Moss Centre was well furnished with woodwork benches, each containing a vice and a selection of hand tools. There were cupboards neatly organised with power tools, additional hand tools and hot glue guns. Around the room, special booths were set out with boxes containing a large variety of paper, cardboard, fabrics, wood, metal and plastic materials, plus a selection of adhesives, string, nails, screws and staplers for joining materials together. Students were informed that if they wanted to use particular materials they could go to the relevant booth, where a member of staff would demonstrate the materials and negotiate their use. These materials were consciously changed or restricted for some of the sessions to encourage the students to apply and extend their skills.

This session aimed to extend previous work carried out during personal and social education lessons at the Birches and to explore familiar and new materials. When they arrived at the workshop at Abraham Moss Centre, the students assembled in a group around the staff where some revision of earlier lessons took place. It was suggested to the students that they could make a puppet to represent a person with whom they would feel safe. They could then present it to the rest of the group on completion. Most of the students were attracted to this idea and for the first fifteen minutes of the session they examined a rod puppet which one of the advisory teachers had brought along. The 'demonstration models' were about 2 - 3 feet in height and had rods which gave their arms quite a range of movements. Initially students were asked to make suggestions about construction while a member of staff moved the puppet, eg "What makes this move? Does the head move? Does the clothing hide any of the parts that make the puppet move?" (Technology AT 4 - Evaluation). They then split into smaller groups and examined the puppets more closely. This gave them further opportunity to discuss and discover more about the construction of the puppet.

The next phase was to choose the materials which would be most suitable for identifying the character they were making, eg uniform for a nurse, doctor, policewoman. Some of the students were able to draw their characters from memory as part of their plan/design which also provided a record of their work. Others needed much more guidance and prompting from members of staff who took photographs to use as a record of the various stages in the process.

<table>
<tr><th colspan="4">'Materials And Their Uses' - Cross -Curricular Aspects</th></tr>
<tr><th>CURRICULUM AREA</th><th>ACTIVITIES</th><th>TECHNOLOGY</th><th>TECHNOLOGY SoA's</th></tr>
<tr><td>ENGLISH</td><td>Discussion about strangers who help us and who we might be safe with.
Discussions during designing and making and recording results.
Making graphics for hats.</td>
<td rowspan="10"><u>AT 1 Level 1</u>
Describing why they might need paper rather than metal to make a hat.

Imagining what it would be like to wear a helmet.

Describing what it would be like if they were lost and who they should ask for help.

<u>AT 2 Level 1</u>
Explaining how they may start to build a structure or vehicle and what features they need.

Drawing pictures of their artefacts and giving reasons for their design.

<u>AT 3 Level 1</u>
Using a various materials to construct puppets.

Using various materials to make vehicles and structures

Using various paper and card to make hats, flowers and masks.

Making choices about which material is likely to be the most suitable.

<u>AT 4 Level 1</u>
Describing to staff and other pupils what they have made and how well they have done.

Describing to others what they like or dislike about what they have made, eg vehicle, puppet or hat.

Discussing and evaluating properties of materials used to make the various artefacts and systems.</td>
<td rowspan="10">"Pupils should be able to:
a) describe to others what they have noticed in familiar surroundings or visualised about imaginary situations.
b) suggest what might be done."

"Pupils should be able to:
a) express their ideas about what they might do to meet an identified need or opportunity."

"Pupils should be able to:
a) use a variety of materials and equipment to make simple things."

"Pupils should be able to:
a) describe to others what they have done and how well they have done it.
b) describe to others what they like and dislike about familiar artefacts, systems and environments."</td></tr>
<tr><td>MATHS</td><td>Sorting balsa wood and other materials into pieces of same length.
Counting number of pieces needed.
Making predictions about whether structure would stand, based on the way it had been constructed.</td></tr>
<tr><td>SCIENCE</td><td>Observing familiar materials at first hand using their senses.
Identifying simple differences, eg hot / cold, rough / smooth.
Knowing that vehicles could be moved by pushing them.</td></tr>
<tr><td>PERSONAL & SOCIAL EDUCATION</td><td>Sharing resources.
Helping each other.
Exploring the idea that there are some strangers who we could trust and ask for help.
Discussing how they might be identified.</td></tr>
<tr><td>GEOGRAPHY</td><td>Finding way round unfamiliar environment.</td></tr>
<tr><td>ART</td><td>Concentrating on aesthetics of artefacts and systems being made, especially hats, flowers, masks and puppets.</td></tr>
<tr><td>DRAMA</td><td>Using hats, masks etc to form basis for drama activity.
Performing play - 'People who help us.'</td></tr>
<tr><td>P.E. / MOVEMENT</td><td>Using masks for drama.
Miming the movement of a character, eg fireman.</td></tr>
<tr><td>HISTORY</td><td>Looking at vehicles from the past and comparing different features, eg lights, horn, wheels, seats etc.
Visiting textile museum and looking at hats from the past.</td></tr>
<tr><td>I.T.</td><td>Use of word processor to make books and record results etc.
Use of computer for graphics.</td></tr>
</table>

Cross-Curricular Links within the Theme of 'Materials & their Uses'

Figure 29

Before they were allowed to start making the puppets, students were reminded of the importance of using tools and materials safely and there was considerable interaction between students and staff concerning the types of tools and materials they had used in school. They were introduced to one new way of joining things together - the hot glue gun - which is a mains powered tool requiring some skill to operate it safely and effectively. After several demonstrations by one of the advisory teachers, the students took turns in joining two pieces of wood with it, closely supervised by the staff. Whilst the hot glue gun was a potentially more dangerous tool than a tube of glue, it extended the students' experiences of the ways in which materials are joined together and offered them greater choice of options. More importantly though, it offered a more accurate, controlled and less messy way of applying glue to a surface. This gave the students greater satisfaction about the presentation of their work.

'"Nurse" Rod Puppet'

Once the materials had been chosen, students began to make their characters and received appropriate intervention from the staff during the whole process. There were several issues/ questions raised and problems to solve, some of which are listed below:

- Which part of the puppet should you start with?
  (>>> Technology AT 4 - Evaluation: AT 1 Identifying a need)

- What colours are in a nurse's/policeman's uniform?
  (>>> Technology AT 2 - Generating a Design)

- What material should the rods be made from?
  (>>> Technology AT 3 - Planning and Making)

- How does the head/hand fit on to the 'body'? Is it successful?
  (>>> Technology AT4 - Evaluation)

- How can the various parts/materials be joined together?
(>>> Technology AT 2 - Generating a Design, AT 3 - Planning and Making)

The students adapted quickly to being in unfamiliar surroundings, got on with their project without prompting and were deeply involved in discussion and negotiation. Most of the puppets were completed by the end of the session and all were taken back to the Birches. Those who had not completed their work were able to select the materials to finish the puppet back at school. Figure 30 provides an example of the first page of a two-page TVEI planning and recording form which is used in the Manchester special schools and shows a summary of the puppet-making session at Abraham Moss Centre. This type of form was used to plan and record every session within the TVEI module 'Technology' which the students were following.

Those students who had not finished their puppets continued the process in other lessons during the week, while the others talked about their characters and how they had made them. Some students were able to write about their visit to the Centre and what they had done. Once all the puppets were completed, the students worked in small groups with members of staff and established what they wanted to say about their policewoman, nurse, doctor, 'lollipop lady', etc. They then presented their work to the rest of the group in turn and talked about an imagined event when this 'person' might be needed. Further expansion of this activity in personal and social education could include the following:

- Pupils' individual statements could be put together, modified and made into a story about all the characters. Some of the students could then present as a play to the rest of the school.

- A more detailed examination of other roles of the various characters, eg police as controlling violence and crime, dog-handlers, motorway police, mounted police.

- Examination of phobias, eg going to the doctor, dentist, nurse and how these might be overcome.

- Exploration of the range of people who wear uniforms or who may be identified by what they wear and examining what they do, eg traffic wardens, soldiers, sailors, railway workers, dentists, post workers, priests, rabbis.

- Visits to the school and talks by a selection of people from the above occupations .

- Visits to see the places where some of the people work and what they do, eg fire station, postal sorting office, dental surgery, hospital, etc.

In Design and Technology, there is clearly further opportunity to examine different ways of making puppets, the materials they are made from, why they might be made in a particular way and how they work. Puppets from around the world could be presented to the students so that they could also learn about various festivals and celebrations when puppets are used. Books or festivals from other countries could create a stimulus for the students to plan their own performance with a puppet they have made. This could then be presented in an end of term performance to pupils and parents. In addition, students could identify the need for a particular method of presentation of the puppets, eg a puppet theatre for glove puppets or a light source for shadow puppets which could be reinforced by visits to 'Punch and Judy' shows, and other puppet performances. They could then design and make their own puppet theatre. Some of the entertainers who often visit special schools would be useful people to get involved with this activity and there may be community artists who could offer further expertise.

A TVEI Enhanced Course

School / College *The Birches*
Course Title *Technology - Materials*
Module Title *Puppet Making*
Date trialled *1991*

No. & time of Lessons
1 lesson at AMC 10.00-11.30am
Enhanced by
English, Maths, Science, PSE

Objectives for the module

To provide students with the opportunity:
to enhance school PSE work (who I feel safe with);
to design and make a puppet;
to explore a variety of materials - fabrics, wood, metal, plastic, paper, etc;
to learn about moving parts / limbs.

Methodology

Revision of PSE lessons.
Demonstration and investigation of a variety of types of puppets and how they are made.
Intervention on an individual / group basis, via discussion and negotiation.
Presentation of possible practical problems, eg what would happen if ...?

Resources

Workshop at Abraham Moss Centre.
Hand tools - hammers, saws, hot glue guns, vices, etc.
Materials - paper, card, wood, metal, fabric, plastic, adhesives, string, sellotape, paint, crayons, pencils, felt tips, etc.

Methods of Assessment

Plans and drawings.
Finished product.
Photographs.

Equal Opportunities: Outcomes

Equal access to all activities, equipment and materials for all students.
Awareness of different skin colours when presenting materials which will represent a person.

TVEI Planning Sheet - 'Puppets'

Figure 30

Sessions 2 & 3 : 'Making a Vehicle'

These two sessions have been described together because the activities in session 3 were generated from the discoveries and identification of needs by a student in session 2. It was intended that during the course of the six weeks the students would be introduced to simple circuitry using batteries, but the work carried out during session 2 provided a natural progression on to the idea of using batteries to power and provide illumination for the vehicles they had made. The staff had planned for the students to begin with a discussion, design their vehicle, make their vehicle and evaluate the finished product which might identify further needs.

Session 2 started with a group discussion of vehicles which are used in everyday life. There were pictures of cars, vans, lorries, coaches etc. and model/toy cars made from various materials which provided a stimulus for developing the conversation. It was suggested to the students that they could make a vehicle using mainly wood for the structure. This it was deliberate attempt to restrict their choice of materials and present them with a considerable challenge which stretched their problem-solving skills in Design and Technology.

'Three of the Student Vehicles'

Safety procedures were revised, then students planned possible designs for the vehicles. Initially, much of the conversation centred upon the structure of the vehicle and the need to strengthen certain parts of it, eg reinforcing the corners of the 'chassis' so that it remained stable. This involved a lot of prompting and intervention from the staff because it was new and unfamiliar to the group. Once the students started to put their plans into practice and make their vehicles, some of them were observed initiating further discussion with staff.

Sakina, who uses a wheelchair and has hemiplaegia, was attempting to saw a piece of wood. In school she tended to depend on her most 'active' hand and easily gave up using the one which had more limited movement. She was having great difficulty in holding her piece of wood with one hand whilst trying to saw it with the other. Sakina obviously needed some help, but the teacher wanted her to try to develop more effective use of both hands - one of her individual objectives throughout the term. She was encouraged to persevere and make demands on her other hand rather than have someone being 'helpful' and doing the task for her. When she eventually succeeded in achieving her goal, she was very excited and several of the students and staff congratulated her on her accomplishment. These activities, together with manipulation of other tools and materials throughout the project, helped Sakina to work on ways of improving her fine motor skills with both hands.

As the making stage progressed, the students were keen to personalise their vehicles and explored ways of designing some 'customising'. Hassan wanted to provide illumination for the front of his vehicle because "cars have lights." This led to an unexpected development in the lesson because he was adamant that it should light up rather than just be symbolic and was not satisfied with metal foil as a substitute. He also commented that he wanted his vehicle to move like one of the battery powered vehicles he was familiar with.

Session 2 was nearly over, however, and there was insufficient time to introduce circuitry to the students. Instead, the whole group and the staff evaluated the vehicles that had been made. Students demonstrated how their vehicles moved and how fast they could go. One of the teachers introduced Hassan's comments about powering the vehicle and everyone was interested in the possibility of making their vehicle move other than by pushing it. It was agreed that the class teacher would provide an introduction to circuitry and batteries back at the school. This ensured continuity and progression from the session at Abraham Moss Centre and also established cross-curricular links with Maths, English and Science as well as Technology.

At the beginning of session 3 the advisory teachers at Abraham Moss Centre talked with the group about the work they had been doing on circuits. The vehicles they had made the previous week were then used to demonstrate the way in which gears could transfer rotation of the motor to driving the wheels. After this demonstration the students were enthusiastic about applying their knowledge to their own vehicles. They formed small groups and began using the motors to make their vehicles move. Hassan reminded the staff that he wanted to make a light for the front of his car. He received some guidance from one of the advisory teachers then found a lamp holder, screws and wire which he secured into the front of his vehicle.

When he had accomplished this task, Hassan was unhappy about making it work by just connecting the wires to the battery. He pointed out that his dad's car did not have wires, but had a switch that made the lights work from inside. This presented further opportunity to experiment with switches and by the end of the session, Hassan's vehicle had a motor which drove the wheels and a light which was operated by a switch. He had also observed that his dad's car started with a key and not by joining wires to the battery. His class teacher explained to him that the key in a car was also a type of switch, but different to the ones the students were using at the Centre. These experiences provided further pupil-led opportunity to extend the work on circuitry and switches within Information Technology (AT5).

Figure 31 illustrates the Design and Technology PoS covered by the activities in these two sessions. It also shows the levels of attainment experienced and achieved by some students across the core subjects of English, Maths and Science, together with Information Technology. Students like Hassan were working within level three, particularly in Science and Design and Technology, while others were working within level one. This emphasises the great potential for individuals with differing attainments to work alongside each other and develop their skills with the appropriate differentiation of activities.

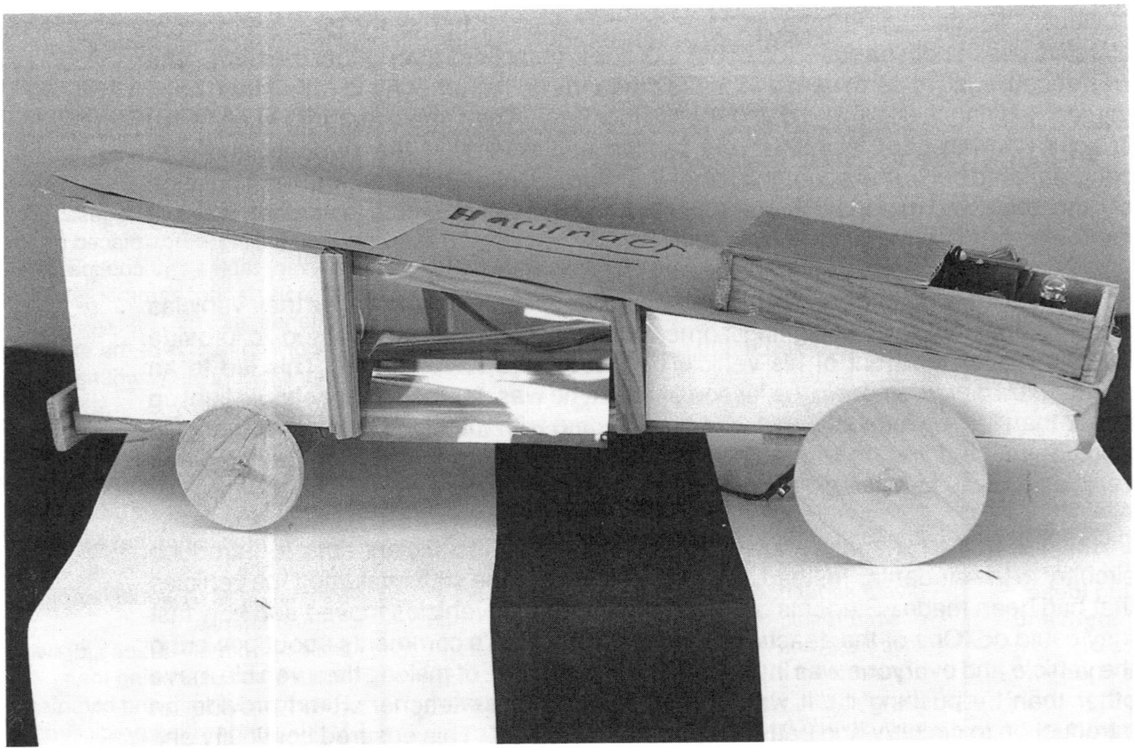

'Hassan's Vehicle'

Figure 32 shows an individual recording sheet for one of the pupils which illustrates the objectives for the pupil and how progress was observed through both the TVEI module and National Curriculum areas of experience.

Session 4 : 'Paper Sculpture'

In the final session observed for this case study the students were introduced to paper sculpture within their school theme of materials and their technology module for TVEI. The workshop at Abraham Moss Centre contained a large selection of different colours, textures and thicknesses of paper and card, as well as various means of cutting and joining pieces together. There was also a computer available set up with a graphics program.

The advisory teachers showed the students and staff from the Birches a selection of hats, flowers and masks made from paper and asked individuals to suggest ways in which certain features were made, eg paper curls and fringes. The teachers then demonstrated several techniques including how they had made the top hat, masks and how the flowers and curls were formed. The students were asked to choose what they wanted to make and most of them had made at least two objects by the end of the session. They were also asked to consider the aesthetics of the finished product during the planning, designing and making process. Not surprisingly, this generated disagreement about what was attractive and what was not. Hassan who was an avid Manchester United supporter did not want to put his name on his hat, but the name and colours of his favourite football team. Some students felt that 'United' were 'rubbish' and preferred to decorate their hat in a more conventional, but nevertheless personal way with flowers and their own name. The names were produced by using a computer graphics program which could also provide some pattern to their design.

Sessions 2 & 3	Design and Technology : 'Making Battery-Powered Vehicles'	
ACTIVITY	**DESIGN & TECHNOLOGY PoS for Levels 1 - 3**	**CORE CURRICULUM AREAS PLUS INFORMATION TECHNOLOGY**
Looking at and discussing pictures of vehicles and model / toy cars. **Deciding type of vehicle to make and what materials would be most appropriate.** **Discussing customising vehicle.** **Selecting material to make vehicle - discussing and exchanging ideas about strength and aesthetics of materials chosen.** **Identifying the need for some form of power to drive the vehicle and illuminate the light.** **Exploring ways of joining pieces of wood and other materials together.** **Tactile exploration of all the materials used.** **Examining the work of others during the making stage and passing comment on the design and function of the vehicle.** **Evaluating their own vehicle and those made by others in the group.** **Exploring making circuits and using switches.** **Recording results by drawings, writing and using word processor.**	**Developing and communicating ideas** *Pupils should be taught to:* *• use imagination and their own experiences, to generate and explore ideas.* *• represent and develop ideas by drawings, models, talking,writing, working with materials.* **Developing and using artefacts, systems and environments** *Pupils should be taught to:* *• recognise that materials can be linked in various ways to make or allow movement.* *• make simple objects for a purpose.* *• know that a system is made of related parts which are combined for a purpose.* *• identify the jobs done by parts of the system.* *• give a sequence of instructions to produce a desired result.* *• recognise and make models of simple structures around them.* *• identify what should be done and the ways in which work should be organised.* *• recognise that control involves making things work as desired.* *• use simple mechanisms to transfer motion.* *• recognise that a source of energy is required to make things work.* *• organise their work, taking account of constraints.* *• use a variety of energy devises.* **Working with materials** *Pupils should be taught to:* *• explore and use a variety of materials to design and make things.* *• recognise that many materials are available and have different characteristics which make them appropriate for different tasks.* *• use materials and equipment safely.* *• investigate the properties of materials in the course of their designing and making.* *• use simple hand tools, and know how to look after them.* *• care for their surroundings.* **Satisfying needs and addressing opportunities** *Pupils should be taught to:* *• realise that resources are limited, and choices must be made.* *• evaluate their finished work against the original intention.* *• ask people about their preferences.* *• talk about what they have learnt and what they might do differently next time.* *• reflect individually and in groups, on how well they went about their work, and whether changes might be needed.*	**ENGLISH** • Involvement in discussion, asking questions, making suggestions. • Recognising words placed on the relevant tables and cupboards, eg tools, fabric, hammer, glue gun, batteries, etc. • Recording results of the experiments / plans by drawings/writing. **MATHS** • Using materials provided for a task. • Talking about own work and asking questions. • Making predictions based on experience. • Estimating what would happen if? • Sorting shapes. • Building with 3D shapes, drawing 2D shapes and describing them. • Recording results and commenting. • Making and testing predictions. • Comparing and ordering objects without measuring. • Examining and making observations about the similarities and differences in a variety of everyday materials. **SCIENCE** • Observing familiar materials and events at first hand, using their senses. • Describing and communicating their observations in groups within their class. • Describing familiar and unfamiliar objects in terms of simple properties, eg size, shape, texture. • Understanding that a complete circuit is needed before an electrical device will work. • Constructing simple circuits. **INFORMATION TECHNOLOGY** • Working with computers. • Talking about ways in which lights and motors responded to switches. • Using computer-generated words or phrases to communicate meaning. • Using information technology for the storage and retrieval of information.

Cross-Curricular Links - 'Making Battery-Powered Vehicles'

Figure 31

The Birches School **Individual Recording Sheet**

CURRICULUM AREA Technology		**NAME** Hassan	
OBJECTIVES • Will use existing knowledge to apply to new situation in discussion, designing & making • Will develop skills in using familiar & new tools • Will evaluate products more objectively, giving reasons.		**DATE SET:** April 1991 **DATE TO REASSESS:** June 1991	
DATE	**ACTIVITY**	**EVALUATION**	**WHOLE CURRICULUM**
15 April	Made puppets at Abraham Moss Centre.	Identified need for puppet to have hat with badge (police).	English, Maths, Science, PSE.
17 April	Wrote about person he felt safe with and how he made his puppet at AMC.	Used word processor and own writing.	English, PSE.
19 April	Presented puppet performance with group at assembly.	Answered questions on how puppet was made and worked.	English, PSE
22 April	Made vehicles from wood at AMC.	Used saw & hot glue gun carefully. Made observations.	Science, Maths, PSE, English.
24 April	Experimented with circuits.	Knew that the wires had to make a circuit for the motor to work.	English, Maths, Science, PSE.
26 April	Experimented with circuits.	Made own circuit with motor and only needed slight prompting.	Science, Maths, PSE, English.
29 April	Made vehicles battery-powered at AMC.	Added light to his vehicle with switch to operate it.	Science, Maths, PSE, English.

EVALUATION Hassan is developing a tremendous talent for model-making and has stretched himself in problem-solving skills particularly. He uses tools safely without having to be reminded and quickly learnt how to operate the new glue gun. He has developed the ability to critically analyse his work instead of thinking everything was "O.K." and can often see ways of improving his work. This is evident in his recording and in the objects he has made. His vehicle is excellent.

Individual Recording Sheet for Part of Topic.

Figure 32

They worked in small groups, sometimes individually with staff, produced their artefacts and then met together as a whole group at the end of the session to evaluate their products. The final discussion generated suggestions about improving the designs, especially as some of the top hats looked fragile. Students also justified why they had chosen a particular colour, shape or form for their objects.

'Paper Hats and Masks by the Birches' Students'

In a previous discussion with the students, it had emerged that they wanted to make a jungle environment in the classroom, using variety of materials. The class staff and students had planned together and negotiated the type of environment they wanted to produce. This topic could create a stimulus for the exploration of wider conservation issues and provide a framework for age-appropriate activities for older students. Some of the activities which could be extended as a result of the students' exploration include the following:

- how paper is made and how it can be recycled.
- an examination of other materials and how they are recycled, eg plastic, aluminium, glass.
- the issue of using softwoods rather than hardwoods and the implications for the rain forests.
- discussion of the types of animals and flowers that might be found in a jungle/ rain forest.
- planning, designing and making paper flowers using knowledge from discussion and skills developed at Abraham Moss Centre. This could include comparison between flowers that grow in different climates.

- designing and making paper flowers and cards decorated with flowers for sale. The money raised could be used for an agreed purpose, eg school fund, a day trip, negotiated by students and staff. Development of this activity into a business enterprise like the one outlined in the Leacroft case study would provide opportunities for further extension of skills and knowledge.

- designing and making animals, birds, flowers, reptiles etc. from a variety of materials including wood, metal, plastic, paper and recycling 'junk'.

- exploring unfamiliar ways of making objects, eg inflatable plastic birds and animals, fur fabric animals, etc. Visiting a manufacturer to look at the process.

- listening to sounds of the jungle.

- evaluating during and after making process.

- writing and drawing about the jungle. Hassan wrote:

  "We have been very busy making our classroom into a jungle. I helped to make some leaves and flowers using different materials. We did some painting, glueing and scrunching."

All the sessions presented opportunities for the students to develop their expertise including the use of Information Technology applied to familiar and unfamiliar circumstances. Some students had used the word processor to write about their experiences which was a familiar way of working. Other experiences were less familiar. In the first session they had used the hot glue gun - a completely new tool which nobody had used before. In the second and third sessions they experimented with switches and circuits and discovered, among other things, that a circuit had to be complete before a motor would rotate or a bulb would light up. In the final session they used a graphics program on the computer - another unfamiliar experience - in order to generate the design and name for the decoration on their hats. Figure 33 illustrates Information Technology experiences for three students of mixed ability.

The Birches School Sample Group Assessment Sheet

SUBJECT / AREA Information Technology

AT	LEVEL	ACTIVITY	PROGRAMME OF STUDY Pupils should be taught:	WORKING WITHIN SoA Pupils should be able to:	OBSERVATIONS
5	1b	HASSAN : to explore the use of batteries and circuits to power his vehicle and make the bulbs light up.	how to give instructions to electronic devices.	talk about ways in which equipment responds to signals.	Hassan needs motivating to find a solution to the problem - making his vehicle move. He will be encouraged to find ways in which the lights would respond if different switches were used.
5	1b	SAKINA : to learn to press a switch that activates a battery - powered vehicle.	how to give instructions to electronic devices.	talk about ways in which equipment responds to signals.	Sakina will be encouraged to take more control, operating different programmes. She will use a single switch plugged into a computer to cause changes and make choices.
5	2a	MARK : to be assisted to video the puppet show. To be encouraged to use an instamatic camera to take photos of the different puppets.	to know that information can be held in a variety of forms.	use information technology for the storage and retrieval of information.	Mark is capable of learning how to operate the school video machine. He can insert the video and rewind it. He will need guidance to locate the "Play" button.

ADDITIONAL IDEAS:

- Encouraging the pupils to use a variety of equipment to demonstrate that switches, keys, dials and knobs will control and activate common household objects, eg radios, televisions, thermostats, electric kettles.
- Use of the computer and word processor to record results of experiments.
- Use of the Touch Screen to activate sounds made by different vehicles.
- Use of graphics program to create a block graph of different vehicles owned by pupils' family.
- Use of a variety of overlays on the concept keyboard to include pictures of different forms of transport together with relevant words.

Figure 33

Conclusion

Technology is essentially a new subject which crosses many curriculum areas. It draws on ideas, concepts, knowledge and skills from other curriculum subjects and uses these to create solutions to practical problems. At the root of good Technology practice is the understanding that Technology capability is NOT about factual content but is about a WAY of learning, a PROCESS in which opportunities for Design and Technology occur spontaneously from children's own need to use their imagination and intuition to creatively solve problems.

Children must be allowed to make mistakes, seek solutions, encouraged to try different ways of solving a problem or satisfying a need. Critics of the problem-solving approach might include the following well-meaning but misguided comments:

> ***"I'll show you."***
>
> ***"No, not like that, what I wanted you to do was ..."***
>
> ***"What's that supposed to be?"***
>
> ***"Look, you join it like this."***
>
> ***"This is how to do it."***
>
> **(Chadwick 1990).**

The four case studies described in this book, support the principles of good Technology practice, in which

- ***"learners should be allowed to become active learners in their own programmes***
- ***a supportive environment should enhance self-confidence and a positive self concept***
- ***pupils should be encouraged to evaluate their own learning"***

(NCC 1989b).

Within the four case studies, we have demonstrated how the opportunities for Design and Technology arise spontaneously and naturally through thematic work, imaginary play, familiar stories and rhymes:

> ***"Technology has its roots in imaginary play, art, craft and thematic work and draws on the programmes of study in other subject areas"***
>
> **(NCC 1990a).**

At the same time, the activities described within the case studies are seen to map fully into the National Curriculum requirements of Technology. The focus of the four case studies is Technology, illustrating the cross-curricular nature of designing, making, testing and evaluating. It also emphasises the links between Technology and English, Maths, Science and Personal and Social Education. In this way our children have so much to gain from Technology which not only embraces the whole curriculum but also makes a vital contribution to Personal and Social Development:

> ***"When Technology is taught through themes, pupils' design and technology activities support their work in other areas. Themes can be chosen which enable the design and technology cycle to be integrated with other work."***
>
> **(NCC 1990a).**

Our experience with pupils with profound and severe learning difficulties is founded on the belief that children learn from the real world around them. Therefore the steps of problem-solving, illustrated within the initial case studies are embedded in learning contexts which are familiar and natural to the children. We have also tried to demonstrate how pupils can progress to working in unfamiliar contexts, thereby allowing for:

> ***"the maximum participation in the Design and Technology attainment targets and programmes of study by all pupils, including those with special educational needs"***
>
> **(NCC 1990a)**

Bibliography

Ackerman, D. and Mount, H. (1991) *Literacy for All.* London: Fulton.

Aherne, P. and Thornber, A. et al. (1990) *Communication for All.* London: Fulton.

Aherne, P. and Thornber, A. et al. (1990) *Mathematics for All.* London: Fulton.

Ashdown, R., Carpenter, B. and Bovair, K. (1991) *The Curriculum Challenge.* London: Falmer Press.

Bell, P. (1990) *Practical Topics for the Primary School. Part 1 - Science and Technology.* Fulwood, Preston: Topical Resources.

Browne, N. (ed.)(1991) *Science and Technology in the Early Years.* Buckingham: Open University Press.

Browne, N. and Ross, C. (1991) *"Girls' Stuff, Boys' Stuff": Young Children Talking and Playing,* in Browne, N. (ed.) Science and Technology in the Early Years. Buckingham: Open University Press.

Byers, R. (1990) *Topics: From Myths to Objectives.* British Journal of Special Education, Volume 17, No.3.

Cassell, B., Lindoe, S. and Skilling, C. (1991) *TVEI and its Relationship to the National Curriculum,* in Ashdown, R., Carpenter, B. and Bovair, K. The Curriculum Challenge. London: Falmer Press.

Chadwick, E. (1990) *Collins Primary Technology-Teachers' Guide for Key Stage 1.* London: Collins Educational.

Department of Education and Science (1967) *Children and Their Primary Schools - A Report of the Central Advisory Council for Education (England).* (Plowden Report). London: HMSO.

Department of Education and Science (1978) *Primary Education in England - A Survey by H.M. Inspectorate of Schools.* London: HMSO.

Department of Education and Science (1987) *Craft, Design and Technology from 5 to 16 - Curriculum Matters 9.* London: HMSO.

Department of Education and Science (1989) *Information Technology from 5 to 16 - Curiculum Matters 15.* London: HMSO.

Department of Education and Science and the Welsh Office (1990) *Technology in the National Curriculum.* London : HMSO.

Fagg, S., Aherne, P., Skelton, S. and Thornber, A. (1990a) *Entitlement for All in Practice.* London: Fulton.

Fagg, S., Skelton, S. et al. (1990b) *Science for All.* London: Fulton.

Hulsegge, J. and Verheuil, A. (1987) *Snoezelen, Another World.* Chesterfield: Rompa.

Lancashire County Council (1990) *Design and Technology - A Practical Approach for Primary Schools.* Pre-publication Version. Lancashire Primary Schools Design and Technology Working Group: Lancashire County Council.

Lever, C. (1990) *National Curriculum Design and Technology for Key Stages 1, 2 and 3.* Stoke-on Trent: Trentham Books.

Manchester City Council (1989) *English National Curriculum - Key Stage 1 Resource Pack.* Manchester City Council Education Department.

Manchester City Council (1990) *Information Technology in the National Curriculum: A Guide for Teachers at Key Satge 1.* Manchester City Council Education Department.

Manchester City Council (1991a) *The 5 -16 Curriculum for Manchester Schools.* Manchester City Council Education Department.

Manchester City Council (1991b) *The 5 to 16 Curriculum: Technology for Manchester Schools.* Draft Copy. Manchester City Council Education Department.

Mc. Kee, D. (1987) *Not now Bernard.* Ingham Yates.

Mills, R. (1991) *Continuity and Progression: Technology with a Purpose.* Article in The Design and Technology Times, Summer 1991. University of Salford: Technology Education Development Unit.

NCC (1989a) *Design and Technology Working Party - Final Report for Key Stage 1.* York: National Curriculum Council.

NCC (1989b) *A Curriculum for All - Special Educational Needs in the National Curriculum.* Curriculum Guidance 2. York: National Curriculum Council.

NCC (1990a) *Non-Statutory Guidance : Design and Technology Capability.* York : National Curriculum Council.

NCC (1990b) *Non-Statutory Guidance: Information Technology Capability.* York: National Curiculum Council.

NCET (1990) *Technology in Support of the National Curriculum for Students with Severe Learning Difficulties: A Resource Pack.* Hatfield Advisory Unit for Microelectronics in Education.

Pring, R. (1989) *The New Curriculum.* London: Cassell Education.

Rose, G. (1980) *Trouble in the Ark.* Ashton Scholastic, by arrangement with Kestrel Books, Penguin Books Ltd.

Schools Examination and Assessment Council (1990) *Learning Through Design and Technology: The APU Model.* London. SEAC.

Wyvil, B. (1991) *Science INSET and Equal Opportunities in the Early Years,* in Browne, N. (ed.) Science and Technology in the Early Years. Buckingham: Open University Press.

References

Reference to the Curricula of the Manchester schools involved in the case studies - Piper Hill, Melland, Leacroft and The Birches has been made throughout this publication.

Birnbaum, I. (1990) *IT in the National Curriculum.* Doncaster: Resource.

East Sussex SLD Schools (1990) *Have We Got a Problem? National Curriculum Guidlines.* Lewes: East Sussex County Council.

National Curriculum Council/BBC (1990) *Technology: 5-11.* York: NCC.

National Curriculum Council (1991) *Science and Pupils with Special Educational Needs: A Workshop Pack for Key Stages 1 and 2.* York: NCC.

National Council for Educational Technology (1990) *Thinking Skills Video: Logo and Children with Severe Learning Difficulties.* Coventry: NCET.

Ouvry, C. and Mitchell, S. (1990) *Make it Simple.* London: Resources for Learning Difficulties.

Potiphar, I.C, Potiphar, P.M.H., Webster, M. and Radford, L. (1989) *Timesavers: The Technology Pack.* Derbyshire: Precise Educational.

Roberts, P. (1991) *Cross Curricular Approaches to Information Technology.* In Ashdown et al. (eds.) above.

Suffolk County Council (1990) *Booklets on Design and Technology Activities at Key Stages 1 and 2.* Suffolk LEA, St. Andrew House, County hall, Ipswich, IP4 1LJ.

Tearle, A. (1991) *Dimensions of Design and Technology.* In Ashdown et al. (eds.)

West Midlands Monitoring Group SLD (1989) *Broadsheet No.2 (Science) and (1990) Broadsheet No.3 (Science).* National Curriculum Monitoring Group (SLD) West Midlands, Universtity of Birmingham, PO Box 363, Birmingham, B15 2TT.

Williams, P. and Jinks, D. (1985) *Design and Technology 5-12.* London, Falmer.

Wren Spinney School (1990) *Modules on Level 1 Science.* Wren Spinney School,

Notes

Notes

Notes